Lecture Notes in Mathematics

Editors:
A. Dold, Heidelberg
F. Takens, Groningen

Springer
Berlin
Heidelberg
New York
Barcelona
Budapest
Hong Kong
London
Milan
Paris
Santa Clara
Singapore
Tokyo

Robert W. Ghrist   Philip J. Holmes
Michael C. Sullivan

Knots and Links in Three-Dimensional Flows

Springer

Authors

Robert W. Ghrist
Department of Mathematics
University of Texas, Austin
Austin, TX 78712, USA
E-mail: ghrist@math.utexas.edu

Michael C. Sullivan
Department of Mathematics
Southern Illinois University at Carbondale
Carbondale, IL 62901, USA
E-mail: msulliva@math.siu.edu

Philip J. Holmes
Program in Applied and Computational Mathematics
and Department of Mechanical and Aerospace Engineering
Princeton University
Princeton, NJ 08618, USA
E-mail: pholmes@rimbaud.princeton.edu

Cataloging-in-Publication Data applied for

Die Deutsche Bibliothek - CIP-Einheitsaufnahme

Ghrist, Robert W.:
Knots and links in three-dimensional flows / Robert W. Ghrist
; Philip J. Holmes ; Michael C. Sullivan. - Berlin ; Heidelberg
; New York ; Barcelona ; Budapest ; Hong Kong ; London ;
Milan ; Paris ; Santa Clara ; Singapore ; Tokyo : Springer, 1997
  (Lecture notes in mathematics ; 1654)
  ISBN 3-540-62628-X
NE: Holmes, Philip:; Sullivan, Michael C.:; GT

Mathematics Subject Classification (1991): 58F25, 57M25, 34C35, 34C25

ISSN 0075-8434
ISBN 3-540-62628-X Springer-Verlag Berlin Heidelberg New York

© Springer-Verlag Berlin Heidelberg 1997
Printed in Germany

The use of general descriptive names, registered names, trademarks, etc. in this
publication does not imply, even in the absence of a specific statement, that such
names are exempt from the relevant protective laws and regulations and therefore
free for general use.

Typesetting: Camera-ready TeX output by the authors
SPIN: 10520337   46/3142-543210 - Printed on acid-free paper

Dedicated to Robert F. Williams,
teacher, colleague, and friend.

Preface

This book was envisaged in the Summer of 1992 at the NATO ASI Conference on Bifurcations of Periodic Orbits in Montreal. From genesis to completion, many new phenomena in the study of knots in dynamical systems have been unearthed (as explored in Chapters 3 through 5).

The authors have benefitted from the support of several institutions, including Cornell's Center for Applied Mathematics, Princeton's Program in Applied and Computational Mathematics, the Institute for Advanced Study, Northwesten University, the University of Texas at Austin, and Southern Illinois University at Carbondale. The authors gratefully acknowledge the support of the National Science Foundation and the Department of Energy.

Finally, we three are indebted to our colleagues who have provided inspiration and pointed out (hopefully most) errors in this text: these include Joan Birman, John Etnyre, John Franks, Jean-Marc Gambaudo, Alec Norton, Dan Silver, and Bob Williams.

Contents

Chapter 0: Introduction

This book concerns knots and links in dynamical systems.

Knot and link theory is an appealing subject. The basic ideas and results may be appreciated intuitively, simply by playing with pieces of string (*e.g.*[11, 1]). Nonetheless, in spite of seafarers' development of sophisticated knots over thousands of years, the mathematical theory of knots began only in the nineteenth century. Its origins lie in Gauss's interest in electromagnetic field lines [67] and in attempts to classify knotted strings in the æther, which Lord Kelvin and others thought might correspond to different chemical elements [176, 174]. It rapidly shed its physical origins and became a cornerstone of low-dimensional topology.

The roots of dynamical systems theory are considerably older and more tangled; they may be found in the *Principia Mathematica* of Isaac Newton and in attempts to model the motions of heavenly bodies. *Ab initio* the subject requires more technical apparatus: the differential and integral calculus, for a start; but at the same time it has kept closer touch with its physical origins. Moreover, in the last hundred years, it too has (re)acquired a strong geometrical flavor. In fact it was in an assault on the (restricted) three body problem of celestial mechanics [145], in response to the prize competition to celebrate the 60th birthday of King Oscar II of Sweden and Norway, that Henri Poincaré essentially invented the modern, geometric theory of dynamical systems. He went on to develop his ideas in considerable detail in *Nouvelles Methodes de la Mécanique Celeste* [146]. Today, following this work, that of the Soviet school, including Pontriagin, Andronov, Kolmogorov, Anosov, and Arnol'd, and of Moser and Smale and their students in the West, the subject has reached a certain maturity. Over the last twenty years, it has escaped from Mathematics Departments into the scientific world at large, and in its somewhat ill-defined incarnations as "chaos theory" and "nonlinear science," the methods and ideas of dynamical systems theory are finding broad application.

The basic world of a dynamical system is its *state space*: a (smooth) manifold, M, which constitutes all possible states of the system, and a mapping or flow defined on M. In one of our principal motivating examples, systems of first order ordinary differential equations (ODEs), the vector field thus specified generates a flow $\phi_t : M \to M, t \in \mathbb{R}$. The general problem tackled by dynamical systems theorists is to describe ϕ_t geometrically, via its action on subsets of M. This implies classification of the asymptotic behaviors of all possible solutions, by finding fixed points, periodic orbits and more exotic recurrent sets, as well as the orbits which flow into and out of them. In many applications ϕ_t also depends on external parameters, and the topological changes or *bifurcations* that occur in M as these parameters are varied, are also of interest. In studying these and related phenomena, one abandons the fruitless search for closed form solutions in terms of elementary or special functions, and seeks instead qualitative information.

Over the past decade, knot theory, once in the inner sanctum of pure mathematics, has been leaking out into other fields through several successful applications. These range from molecular biology, involving topological structures of closed DNA strands [173], to physics, led by surprising connections with statistical mechanics [99] and quantum field theory [197, 14]. Likewise, over the past ten to fifteen years, several attempts have been made to draw knot theory and dynamical systems closer together. The key idea is simple: a closed (periodic) orbit in a three-dimensional flow is an embedding of the circle, S^1, into the three-manifold that constitutes the state space of the system, hence it is a knot. Similarly, a finite collection of periodic orbits defines a link.

Several natural questions immediately arise, directed at the following goal: given a flow, perhaps generated by the vector field of a specific ODE, describe the knot and link types to be found among its periodic orbits. Do nontrivial knots occur? How many distinct knot types are represented? How many of each type? Do well-known families, such as torus knots, algebraic knots, or rational tangles, appear in particular cases? In any cases? Are there "new" families of knots and links which arise naturally in certain flows? Do Hamiltonian and other systems with conservation laws or symmetries support preferred families of links? Do "chaotic" flows contain inherently richer knotting than simple (Morse-Smale) flows? Indeed, how complicated can things get? – is there a single ODE among whose periodic orbits can be found representatives of *all* knots and links? Such questions might occur to topologists. Indeed, it was R.F. Williams, in the context of a seminar on turbulence conducted in the Mathematics Department at Berkeley in 1976, who first conjectured that nontrivial knotting occurs in a well-known set of ODEs called the Lorenz equations [193].

Dynamicists, in contrast, might seek to use knot and link invariants to describe periodic orbits and so help them better understand the underlying ODEs. In a parametrised family of flows, for example, one can observe sequences of bifurcations in which a simple invariant set containing, say, one or two periodic orbits, "grows" into a chaotic set of great complexity, containing a countable infinity of periodic orbits. In many cases, the periodic orbits are dense in the set of interest; sometimes that set is a so-called strange attractor. The existence-uniqueness theorem for solutions of ODEs implies that, as periodic orbits deform under parameter variation, they cannot intersect or pass through one another. Knot and link types therefore provide topological invariants which may be attached to families of periodic orbits. Can such invariants be used to identify orbit genealogies – to trace the bifurcation sequences in which they arose? (A favorite problem is to describe bifurcation sequences in the two-parameter family of maps introduced by Hénon [83], which provides a model for Smale's famous horseshoe map.) Can operations in which new knots are created from old, such as composition and cabling, be associated with specific local bifurcations? Is the complexity of knotting related to other measures of dynamical complexity, such as topological entropy? Does knot theory provide finer invariants than entropy for the classification of flows?

Of course, since periodic orbits form knots only in *three*-dimensional flows, applications to dynamical systems in general are severely limited. Nonetheless, many of the rich and wonderful behaviors that currently engage dynamicists are already manifest in three dimensions, and so it seems well worth applying whatever tools we can to this case. In any event, we hope the reader will find the subject as beautiful, and attractive, as we do.

0.1 The contents of this volume

This book attempts to bring together two largely disparate and well developed fields, which have thus far only met in the pages of specialised research journals. As such, it cannot substitute for a proper course or text in either field. Chapter 1, to follow immediately, provides a rapid review of the principal aspects of knot theory and dynamical systems theory required for the remainder of the book. In Chapter 2 we develop the major tool which allows us to pass back and forth between hyperbolic flows and knots: the *template*. This was introduced (under the name "knot holder") over twelve years ago in two papers of Birman and Williams [23, 24]. In dynamical systems it is common to use *Poincaré* or *return maps* to reduce a flow to a mapping on a manifold of one lower dimension. While Poincaré maps preserve certain periodic orbit data, information on how the orbits are embedded in the flow is lost. The template preserves that information, and likewise reduces dimension. In Chapter 2 we develop a host of related tools: subtemplates, template inflations and renormalisations, and the symbolic language which allows us to manipulate templates and explore relations among them. We also introduce some of the particular (families of) templates which will concern us later.

Equipped with our basic tools, in Chapter 3 we obtain some general results on template knots and links, including the facts that, while specific templates may not contain *all* knots and links, every template contains infinitely many distinct knot types. We then describe a *universal* template, which *does* contain all (tame) knots and links, and which, moreover, arises rather naturally in certain classes of structurally stable three dimensional flows. In the final section, we explore the "embedding problem:" the question of which templates can be embedded in other templates. By considering isotopic embeddings, we are able to recognise universal templates hidden in ostensibly simpler ones.

The fourth chapter concerns bifurcations and knots, and directly addresses the kinds of dynamical systems questions raised in our opening paragraphs. In particular we focus on specific templates related to the Hénon mapping and the creation of horseshoes. Here, in contrast to the limitless riches of Chapter 3, there are severe restrictions on links (all crossings are of one sign), which lead to uniqueness results and order relations on orbit creation in local bifurcations. We also explore knot types born in certain global or homoclinic bifurcations, by lifting the contrast between dynamically simple and dynamically complex

bifurcations to the knot-theoretic level. In so doing, we derive a rather general
set of sufficient conditions for a third-order ODE to support all links as periodic
orbits.

Chapter 5 returns to basic template theory and presents the current state
of affairs in template classification and invariant theory. We commence with
a discussion of what a sensible definition of template equivalence should be,
based on intuition developed in Chapters 3 and 4, and continue with a primitive
but useful invariant: a zeta-function for a restricted class of templates. This
will be seen to relate nicely to the underlying symbolic dynamics, yielding an
easily-computed invariant which encodes "twisting" information in the compact
package of a rational function.

Chapter 6 is comprised of a short list of concluding remarks and open pro-
belms that pertain to template theory and its applications.

Throughout Chapters 2-5 we strive to present, for the first time, a fairly
complete picture of the theory of templates. As such, we include key results
of Franks, Birman, Williams and others, although we focus primarily on our
own work, relegating to an appendix some related work beyond the immediate
scope of this monograph. Accordingly, Appendix A contains brief reviews of
work by Morgan, Wada, and others on nonsingular Morse-Smale flows on three-
manifolds, which contain only limited classes of knots. This is then contrasted
with the work of Franks and work in progress by Sullivan on nonsingular Smale
flows on the three-sphere.

Despite the title, we in no way claim to include every major result in the
overlap of dynamics and knot theory. In particular, there is a natural dichotomy
between knots arising from suspended surface homeomorphisms and those arising as closed orbits in flows on three-manifolds: this text focuses on the latter
situation. The forthcoming book by P. Boyland and T. Hall [31] deals with the
former — there is a great deal of beautiful work being done in this area: Nielsen
theory and "braid types" for surface automorphisms [30, 29]. In addition, knot
theory intersects with dynamics in examining problems of integrable Hamiltonian systems [50], the existence of minimal flows on three-manifolds [79] and
contact geometry [45]. Finally, analogues of knotting and linking for nonperi-
odic, minimal orbits [15, 116] and "asymptotic" linking of orbits [64, 62] are very
exciting, particularly since there are applications to magnetohydrodynamics [7]
and fluid mechanics [129].

Chapter 1: Prerequisites

Before introducing the tools for examining knotted periodic orbits in flows, we provide a concise review of relevant definitions, ideas, and results from the topological theory of knots and links and the dynamical theory of flows in three dimensions. This provides a language for describing phenomena, as in: *a period-doubling bifurcation gives rise to a (2, n) cabling.*

Our treatment of both of these (large) bodies of theory is necessarily brief; we wish merely to describe the main ideas to be used in subsequent chapters. Several good references exist for these growing fields. Standard texts for the theory of knots and links includes the books by Rolfsen [154], Burde and Zieschang [33], and Kauffman [101]. In the theory of dynamical systems, a wealth of good books can be found, including those by Robinson [153], Shub [162], Arnold [6], and Bowen [26]. Devaney's book [41] is a good introductory text on iterated mappings. A more applied viewpoint can be found in the texts by Guckenheimer and Holmes [76] or Arrowsmith and Place [9].

1.1 The theory of knots and links

Given a piece of string, one may tie it up into all sorts of complicated knots. Nevertheless, as long as the ends are free, the mess may be untied completely (though in practice this may be frustrating!). If one should join the two free ends of the string together, then (intuitively) a knotted loop remains knotted no matter how one tries to undo it. This is the idea behind knot theory.

1.1.1 Basic definitions

Definition 1.1.1 A *knot* is an embedding $K : S^1 \hookrightarrow S^3$ of a 1-sphere into the 3-sphere. A *link* $L : \coprod S^1 \hookrightarrow S^3$ is a disjoint, finite collection of knots.

The three-sphere S^3 is defined as the unit sphere in $\mathbb{R}^4$. The reader who is uncomfortable with S^3 may replace it by $\mathbb{R}^3$ without loss, since S^3 can be considered as $\mathbb{R}^3$ with an additional "point at infinity." The simplest knot is the *unknot*, pictured in Figure 1.1(a). An unknot is any embedding of S^1 in S^3 whose image is the boundary of an embedded disc $D^2 \subset S^3$. The next "simplest" knots[1] are the *trefoil knot* and the *figure-eight knot* depicted in Figure 1.1. We will usually consider knots and links which are *oriented*, as depicted by an

---

[1] The first knot theorists tabulated knots according to the minimal number of crossings in a planar projection. In these tables (see [154] or [33]) the knots of Figure 1.1 are simplest: *i.e.*, they have the fewest possible number of crossings. Other notions of "simplicity" are of course possible [115].

Figure 1.1: (a) the unknot; (b) the trefoil knot; (c) the figure–eight knot.

arrow along the knot in a diagram. Given some regular (i.e., transverse) planar representation of an oriented knot or link, each crossing point has an induced orientation, given by the convention of Figure 1.2. While our convention is opposite that which is standard in knot theory, it has prevailed in the study of knots in dynamical systems [23, 24, 93, 88, 89, 70].

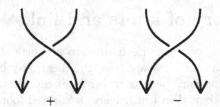

Figure 1.2: Sign convention for crossings.

The fundamental problem of knot (link) theory is the following: when are two knots (links) the same? In knot theory, the notion of "sameness" is constructed to match our intuition of deforming loops of knotted string.

Definition 1.1.2 Two knots K and $\tilde{K}$ are *ambient isotopic* if there exists a continuous one-parameter family h_t of homeomorphisms of S^3 such that h_0 is the identity map and $h_1 \circ K = \tilde{K}$.

Remark 1.1.3 The natural analogue of Definition 1.1.2 holds for embeddings of spaces in S^3 other than S^1, *e.g.*, surfaces and solids. When working with knots and links in S^3, it is common to refer to ambient isotopic knots as being *isotopic*, even though *isotopy* is technically a weaker equivalence when working with noncompact spaces [33]. We use the terms interchangeably to denote the equivalence of Definition 1.1.2.

Unless specified explicitly, the term "knot" may refer to either the actual embedding, or the image of the embedding, or the entire isotopy class of embeddings. We will formulate most of the theory in terms of knots — generalizations to links are automatic.

Given Definition 1.1.2, the fundamental problem of knot theory can be stated as follows:

Problem 1.1.4 When are two knots isotopic?

One of the first triumphs of knot theory was a reformulation of Problem 1.1.4 from a global-topological problem to a local-combinatorial one due to Reidemeister [149]. Given a knot or link, consider all its *presentations*; that is, planar projections with overcrossings and undercrossings marked as in Figure 1.1. Any presentation may always be chosen such that it is *regular*, having only transverse double-points.

Theorem 1.1.5 (Reidemeister [149]) *Two regular presentations correspond to isotopic links if and only if the diagrams are related by isotopy (fixing the crossing points) and by a finite sequence of the three Reidemeister moves, given in Figure 1.3.*

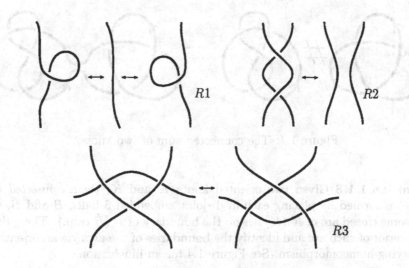

Figure 1.3: The three Reidemeister moves: R1, R2, R3.

Even with Theorem 1.1.5, Problem 1.1.4 is very difficult to solve; however, restricted versions of this problem have clean solutions.

Consider the class of *torus knots*: that is, knots which lie on a torus $T^2 = S^1 \times S^1 \subset S^3$, where each S^1 is unknotted. These knots are described by their winding number in the meridional and longitudinal directions. A type (m, n) torus knot (m and n relatively prime positive integers) is a simple closed curve on T^2 which winds about the longitudinal direction m times and about the meridional direction n times [154, 33].

Example 1.1.6 The trefoil knot of Figure 1.1(b) is a (2,3) torus knot.

The family of torus knots is well-understood; in particular, we have:

Proposition 1.1.7 ([154, 33]) *Torus knots of type (m,n) and (m',n') are isotopic if and only if $m = m'$ and $n = n'$ (or, equivalently, $m = n'$ and $n = m'$).*

1.1.2   New knots from old

One possible method for building and classifying knots is to begin with a simple family (*e.g.*, the torus knots) and combine its members in various ways. Given two knots, there are certain constructions for creating a new knot: we shall consider two such operations which also have dynamical interpretations.

Connected sums

The first operation we consider is a form of "multiplication" for knots called, oddly enough, the *sum*.

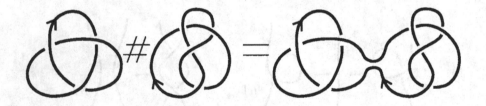

Figure 1.4: The connected sum of two knots.

Definition 1.1.8 Given two oriented knots K and $\tilde{K}$, their *connected sum*, $K \# \tilde{K}$, is formed by placing each in disjoint embedded 3-balls, B and $\tilde{B}$, such that some closed arc of K ($\tilde{K}$) lies on the boundary of B ($\tilde{B}$ resp.). Then, delete the interior of each arc and identify the boundaries of the arcs via an orientation preserving homeomorphism. See Figure 1.4 for an illustration.

Remark 1.1.9 In Definition 1.1.8, the choice of balls and arcs does not affect the connected sum. This operation is commutative and associative, but is not a group operation due to the lack of inverses [154].

If a knot can be decomposed into the connected sum of two or more nontrivial knots, it is said to be *composite*, else it is *prime*. The torus knots, for example, are prime (a nice proof can be found in [33, pp. 92-93]. A classical theorem due to Schubert states that every knot has a unique prime factorization as the connected sum of prime knots. R. F. Williams and M. Sullivan have explored the presence of prime decompositions of periodic orbits of flows [195, 169].

Companions and satellites

If one thinks of the connected sum as a form of multiplication on the space of all knots (complete with prime factorization as with the integers), the operation of taking *satellites* is akin to taking powers. Let $V \equiv D^2 \times S^1$ be a solid torus which sits in S^3 in the standard way. Let K be a knot essentially embedded in V, i.e., K is not contained in any 3-ball $B \subset V$. Let $\tilde{K}$ be an arbitrary knot and $N_{\tilde{K}}$ a tubular neighborhood of this knot in S^3. A homeomorphism $h : V \to N_{\tilde{K}}$ is said to be *faithful* if it takes the longitude of ∂V to a longitude of $\partial N_{\tilde{K}}$ which is homologically trivial (it bounds a surface) in the complement $S^3 \setminus N_{\tilde{K}}$.

Definition 1.1.10 The image of K under a faithful homeomorphism h is a *satellite knot* with *companion* $\tilde{K}$ and *pattern* (K, V): see Figure 1.5. If K isotopes to a subset of $\partial V \equiv T^2$, then K is a (p, q) torus knot and $h(K)$ is said to be the (p, q) *cable* of $\tilde{K}$.

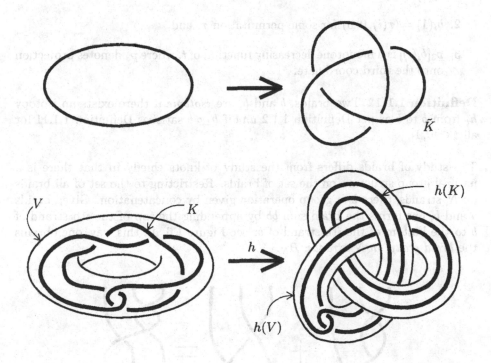

Figure 1.5: A companion $(\tilde{K})$ and a satellite $(h(K))$ knot.

If we take $\tilde{K}$ to be the unknot, a (p_1, q_1) cable of $\tilde{K}$ is a (p_1, q_1) torus knot. If we build a (p_2, q_2) cable of this torus knot, we obtain a new knot. By continuing this procedure, with (p_i, q_i) cablings at each step, one produces an *iterated torus knot* of type $\{(p_i, q_i)\}_{i=1}^n$. Alternatively, we say that the set of knots generated from the unknot by the operation of cabling is called the set of *iterated torus knots.* Following Fomenko and Nguyen [50], we will denote the set of knots generated from the unknot by the operations of cabling and connected sum the set of *generalized iterated torus knots.* Both of these families of knots arise naturally in a dynamical context as shown in Appendix A.

1.1.3   Braid theory

Knot and link theory studies embeddings of circles in S^3. With some slight restrictions on the range of the embeddings, one can also embed arcs in topologically distinct ways. *Braid theory* studies these phenomena (see [19, 81]):

Definition 1.1.11 Given N a positive integer, a *braid on N strands* is a collection $b = \{b_i\}_1^N$ of N disjoint embeddings of the interval $[0, 1]$ into Euclidean $\mathbb{R}^3$ such that for each i,

1. $b_i(0) = (i, 0, 1)$;

2. $b_i(1) = (\tau(i), 0, 0)$ for some permutation τ; and

3. $p_3[b_i(t)]$ is a monotone decreasing function of t, where p_3 denotes projection onto the third coordinate.

Definition 1.1.12 Two braids, b and $\tilde{b}$, are *isotopic* if there exists an isotopy h_t from b to $\tilde{b}$ as per Definition 1.1.2 *and* if $h_t \circ b$ satisfies Definition 1.1.11 for all $t \in [0, 1]$.

The study of braids differs from the study of knots chiefly in that there is a natural group structure on the set of braids. Restricting to the set of all braids on N strands, there is a group operation given by concatenation. Given braids b and $\tilde{b}$, one forms the braid sum $b\tilde{b}$ by appending the top of the ith strand of $\tilde{b}$ to the bottom of the ith strand of b: see Figure 1.6. In this way, one obtains the *braid group on N strands*, B_N.

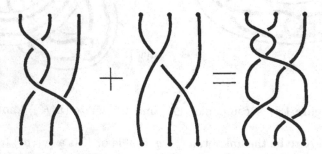

Figure 1.6: The sum operation on the braid group B_3: $\sigma_1^2 \sigma_2$ concatenated with $\sigma_1^{-1} \sigma_2$ equals $\sigma_1^2 \sigma_2 \sigma_1^{-1} \sigma_2$.

The standard generators for B_N are denoted $\{\sigma_i : i = 1...(N-1)\}$ and are given geometrically as the crossing of the ith strand over the $(i+1)$st strand, as depicted in Figure 1.7. The presentation for B_N under these generators was given by Artin [10] to be the following:

$$B_N = \left\langle \sigma_1, \sigma_2, \ldots, \sigma_{N-1} : \begin{array}{cc} \sigma_i \sigma_j = \sigma_j \sigma_i & , \quad |i-j| > 1 \\ \sigma_i \sigma_{i+1} \sigma_i = \sigma_{i+1} \sigma_i \sigma_{i+1} & , \quad i < N-1 \end{array} \right\rangle. \quad (1.1)$$

The relations for this presentation are illustrated in Figure 1.8.

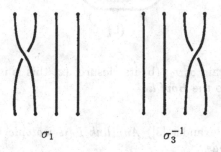

$$\sigma_1 \qquad\qquad\qquad \sigma_3^{-1}$$

Figure 1.7: Examples of generators for the braid group B_4.

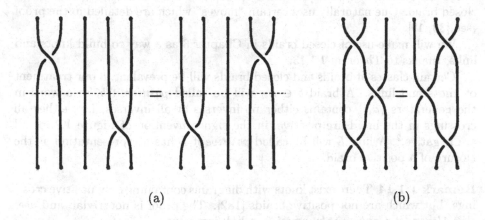

(a)   (b)

Figure 1.8: Relations for the braid group B_N: (a) $\sigma_i \sigma_j = \sigma_j \sigma_i$ for $|i-j| > 1$; (b) $\sigma_i \sigma_{i+1} \sigma_i = \sigma_{i+1} \sigma_i \sigma_{i+1}$ for $i < N-1$.

A relationship between braid theory and link theory is established by a simple operation on braids known as *closure*. Given a braid b, one forms a *closed braid*, $\bar{b}$, by connecting the top and the bottom of each strand of b in the obvious fashion: see Figure 1.9. The question of the extent to which closed braids represent knots and links was answered by Alexander [3]:

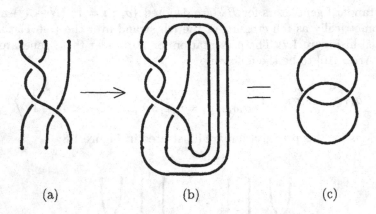

(a)   (b)   (c)

Figure 1.9: (a) the braid $\sigma_1^2\sigma_2$; (b) its closure; (c) this is isotopic (via the first Reidemeister move) to the *Hopf link*.

Theorem 1.1.13 (Alexander [3]) *Any link L is isotopic to a closed braid on some number of strands.*

To understand the proof of Theorem 1.1.13, the reader is encouraged to isotope a closed piece of string into a closed braid: choose a provisional *braid axis*, about which the strands should revolve, and then try to maneuver the strands into a closed braid. One naturally uses certain "moves" which are detailed in the proof (see [33, 19]).

We will make use of closed braids in Chapter 3 as a way to build knots and links, thanks to Theorem 1.1.13.

Certain classes of braids and closed braids will be prevalent in our treatment of knots and links. A braid $b \in B_n$ will be called *positive* if b, as a word in the generators $\{\sigma_i\}$, contains either no inverses or all inverses, *i.e.*, either all crossings in the braid are positive, in the sign convention of Figure 1.2, or all are negative.[2] A link L will be called *positive* if L has a representation as the closure of a positive braid.

Remark 1.1.14 There exist knots with diagrams containing only positive crossings, but which are not positive braids [182]. The proof is nontrivial, and uses the *Alexander-Conway polynomial* — a link invariant.

1.1.4   Numerical invariants

The equivalence problem (Problem 1.1.4) for knots and links is extremely difficult and has not yet been solved in a computationally reasonable manner. However,

---

[2]The term *positive* is used in both cases, either all positive crossings or all negative crossings. We find this confusing and would prefer the term *uniform*; however, we yield to the common practice in the remainder of this work.

many advances have been made through the use of *algebraic invariants* (see [101, 99, 20, 21, 94, 59] for examples). Here we merely describe some simpler, classical, *numerical invariants*, which will suffice for out purposes.

A *numerical invariant* is a well-defined function from link equivalence classes to the integers. For example, the function which maps a link L to the number of its components $\mu(L)$ is obviously invariant under isotopy, and hence defines a numerical invariant. However, this invariant has rather poor eyesight, since it does not distinguish different n-component links.

Consider a link L of two components, K and $\tilde{K}$. There is a well-defined notion of how "entwined" K and $\tilde{K}$ are, encoded in the *linking number*, $\ell k(K, \tilde{K}) \in \mathbb{Z}$. There are numerous ways to define linking number [154], the simplest of which involves a presentation of the link (recall Theorem 1.1.5). For an oriented link, one can label each crossing of a regular link presentation with an integer ± 1, as per the convention of Figure 1.2.

Definition 1.1.15 Given two knots K and $\tilde{K}$, the *linking number*, $\ell k(K, \tilde{K})$, is given as half the sum of the signs over all crossings of K with $\tilde{K}$,

$$\ell k(K, \tilde{K}) = \frac{1}{2} \sum_{K \cap \tilde{K}} \epsilon_i, \qquad (1.2)$$

where $\epsilon_i = \pm 1$ is the sign of the ith crossing and $K \cap \tilde{K}$ denotes the crossings of K and $\tilde{K}$ in some regular presentation.

Lemma 1.1.16 *Linking number is a link isotopy invariant.*

Proof: By Theorem 1.1.5, isotopy is generated by the Reidemeister moves of Figure 1.3. It is easy to verify that linking number does not change under these local moves.   □

The linking number $\ell k(K, \tilde{K})$ is related to the intuitive notion of linking. For example, define a *separable* link to be one for which there exists a smooth embedded 2-sphere S^2 in S^3 which separates one (or more) component(s) of L from the remainder of L. Any two separated components of a link are said to be *unlinked*, and, indeed, their linking number must be zero, since there exists a presentation for the link in which the components do not cross at all. We note, however, that it does not follow that two knots with linking number zero are necessarily separated: see the *Whitehead link* of Figure 2.16 for a classical example.

One of the most important numerical invariants is the *genus* of a link. Recall that closed orientable surfaces are classified by *genus*, or the number of handles in a handlebody decomposition. Similarly, the genus of any surface with boundary is defined as the genus of the surface obtained by abstractly gluing in a disc along each boundary component.

Definition 1.1.17 Given a link L, the *genus*, $g(L)$, is defined as the minimum genus over all orientable surfaces S which span L: that is, $\partial S = L$, where ∂S

is the oriented boundary. A spanning surface of minimal genus is known as a *Seifert surface*.

Genus is by definition an invariant. Since by definition a knot in S^3 bounds a disk if and only if it is the unknot, then among knots, only the unknot may have genus zero.

There are numerous formulae available for computing genera of links. We include one, due to Birman and Williams [23], following work of Stallings [167], which will be particularly useful in later chapters.

Theorem 1.1.18 (Birman and Williams [24]) *Let L be a non-separable link of μ components, presented as a closed positive braid on N strands, with c crossings. Then $g(L)$, the genus of L, is given as*

$$g(L) = \frac{c - N - \mu}{2} + 1. \tag{1.3}$$

Example 1.1.19 In Figure 1.10(a), we show the trefoil knot along with a spanning surface. An Euler number calculation reveals that the surface is a punctured torus whose genus is one. By using Equation (1.3) on the (positive) braid representation in Figure 1.10(b), we get $\mu = 1$, $c = +3$, and $N = 2$; hence, the genus is one, and the surface of part (a) is actually the Seifert surface. This proves that the trefoil is indeed knotted.

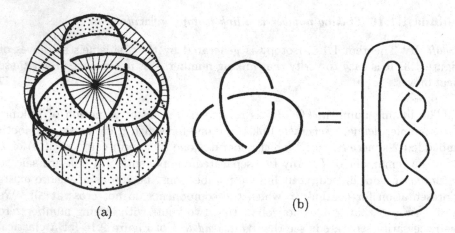

(a)   (b)

Figure 1.10: (a) A spanning surface for the trefoil knot; (b) a positive braid presentation

Example 1.1.20 We may extend the idea of Example 1.1.19 to compute a general formula for the genus of a torus knot. For K a (m, n) torus knot with $m > n$, we present a presentation of K as a positive braid in Figure 1.11: there are m strands on a cylinder (the logitudinal direction), n of which twist around the back (the meridional direction). The closure of this braid is K. It is an exercise for the reader to count the crossings in this illustration and, using Equation (1.3), compute the genus of K to be:

$$g(K) = \frac{(m-1)(n-1)}{2}. \tag{1.4}$$

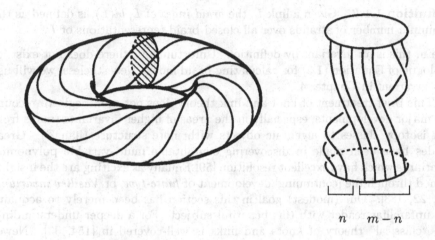

Figure 1.11: The (m, n) torus knot as a positive braid on m strands.

Exercise 1.1.21 The figure-8 knot has genus one and braid word $\sigma_1 \sigma_2^{-1} \sigma_1 \sigma_2^{-1}$. Show that it cannot be presented as a positive braid. Hint: use induction on the number of strands.

Solution: Clearly, one or two strands will not suffice. For three strands, there must be precisely four crossings to ensure genus one. Show that any positive braid with four crossings is either a trefoil or a link with more than one component. For $N > 3$ strands, $c = N + 1$, and, given a positive braid on N strands with $N + 1$ crossings, there must be one braid generator that is only used once. Thus, by "flipping" as in the first Reidemeister move, one can reduce the number of strands while retaining positivity, and thus obtain a counter example on $N - 1$ strands. $\square$

The condition of having a positive closed braid is crucial to Theorem 1.1.18. For non-positive (or *mixed*) braids, there exists an extension of Theorem 1.1.18 due to Bennequin [17], who derived a lower bound for genera of closed braids given the same data as in Theorem 1.1.18:[3]

---

[3]The upper bound follows from direct construction.

Theorem 1.1.22 (Bennequin [17]) *Let L be a nonseparable link of μ components, presented as a closed braid on N strands, with c_+ (c_-) crossings in the positive (negative) sense. Then $g(L)$, the genus of L, is bounded as follows:*

$$\frac{|c_+ - c_-| - N - \mu}{2} + 1 \leq g(L) \leq \frac{|c_+ + c_-| - N - \mu}{2} + 1. \qquad (1.5)$$

There are numerous other classical numerical invariants for knots and links: we mention one last example for future reference.

Definition 1.1.23 Given a link L, the *braid index* of L, $bi(L)$, is defined as the minimum number of strands over all closed braid representations of L.

Again, this is an invariant by definition. Unfortunately, there does not exist an analogue of Equation (1.3) for calculating braid index. Nevertheless, we will use this invariant in Chapter 4.

This brief treatment of knot and link theory does not even begin to recount the major developments, especially in the areas of higher invariants (maps from link isotopy classes to algebraic objects with more structure than $\mathbb{Z}$). Great strides have been made in discovering computable multi-variable polynomial invariants which have excellent resolution [59].Equally as exciting are the insights gained through the [continuing] development of *finite-type*, or *Vasiliev invariants* [21, 22, 183]. Our (modest) goal in this section has been merely to acquaint the unfamiliar reader with this beautiful subject. For a deeper understanding, the "classical" theory of knots and links is well-covered in [154, 33]. Newer perspectives can be found in [21] and the references therein. Braid theory is covered in [19], with more recent progress reported in [20].

1.2   The theory of dynamical systems

Topology is the study of continuous maps between topological spaces: $f : X \to Y$. In the case where $f : X \to X$, one is easily persuaded to consider iterated points or *orbits* of f. Dynamics seeks to understand asymptotic properties of orbits, be they orbits of maps ($\mathbb{Z}$-actions) or of flows ($\mathbb{R}$-actions). In the case of flows on 3-manifolds, we will consider the topological properties of closed orbits as knots and links. But in order to proceed, we will need a certain amount of terminology and theory for both maps and flows.

1.2.1   Basic definitions

Discrete dynamics

Although dynamical systems originated in questions about continuous-time dynamics (in celestial mechanics; see, for example, the historical account in [43]), much of the theory was developed first for maps, as it is somewhat simpler in this case. Thus, in this section, we assume $f : M \to M$ is a diffeomorphism of an n-manifold M. The *orbit*, $o(x)$, of a point $x \in M$ is defined as the set of iterates $\{f^k(x) : k \in \mathbb{Z}\}$.

Remark 1.2.1 Although we state the results for diffeomorphisms, much of the theory goes through for smooth noninvertible maps, for which one works with the orbits $\{f^k(x) : k \in \mathbb{N}\}$. The case of one-dimensional noninvertible maps will be of particular concern in Section 1.2.3, and subsequently in the study of semiflows on templates.

There are two primary problems associated to the dynamics of maps. The first is the equivalence problem (*cf.* Problem 1.1.4):

Definition 1.2.2 Two diffeomorphisms $f : M \to M$ and $\tilde{f} : N \to N$ are *conjugate* if there exists a homeomorphism $h : M \to N$ such that the following diagram commutes:

$$
\begin{array}{ccc}
M & \xrightarrow{f} & M \\
h \downarrow & & \downarrow h \\
N & \xrightarrow{\tilde{f}} & N
\end{array}
\tag{1.6}
$$

Problem 1.2.3 *When are two diffeomorphisms conjugate?*

The second principal problem of dynamics concerns stability: when are *all* "nearby" maps equivalent?

Definition 1.2.4 A diffeomorphisms $f : M \to M$ is *structurally stable* if all diffeomorphisms in a sufficiently small neighborhood of f in $C^1(M)$ are conjugate to f.

Problem 1.2.5 *When is a map structurally stable?*

Problems 1.2.3 and 1.2.5 are relevant, not only to the study of maps and flows (to be discussed below), but also to the physical processes that are frequently modeled by such systems. They are large problems, whose study has spawned a number of important results and perspectives.

We begin by breaking the problem down. An *invariant set* of f is a subset $\Lambda \subset M$ such that $f(\Lambda) = \Lambda$. An *equilibrium*, or *fixed point* for f is a one-point invariant set. Understanding of the behavior on an invariant set Λ is greatly facilitated if the action of f on Λ can be decomposed into uniformly expanding and contracting pieces. This is the kernel of the notion of *hyperbolicity*.

Definition 1.2.6 An invariant set $\Lambda \subset M$ for a map $f : M \to M$ is *hyperbolic* if there exists a continuous f-invariant splitting of the tangent bundle TM_Λ into *stable* and *unstable bundles* $E_\Lambda^s \oplus E_\Lambda^u$ with

$$
\begin{aligned}
\|Df^n(v)\| &\leq C\lambda^{-n}\|v\| \quad \forall\, v \in E_\Lambda^s, \quad \forall\, n > 0, \\
\|Df^{-n}(v)\| &\leq C\lambda^{-n}\|v\| \quad \forall\, v \in E_\Lambda^u, \quad \forall\, n > 0,
\end{aligned}
\tag{1.7}
$$

for some fixed $C > 0, \lambda > 1$.

If f is hyperbolic on all of M, we say that f is *Anosov*. Given a hyperbolic structure on an invariant set, the dynamics and stability of orbits on that piece are well-understood, as we now describe.

Example 1.2.7 (the toral Anosov map) Consider the linear map $f : \mathbb{R}^2 \to \mathbb{R}^2$ given by

$$f : \begin{pmatrix} x \\ y \end{pmatrix} \mapsto \begin{bmatrix} 2 & 1 \\ 1 & 1 \end{bmatrix} \begin{pmatrix} x \\ y \end{pmatrix} = \begin{pmatrix} 2x + y \\ x + y \end{pmatrix}. \tag{1.8}$$

The point $(0,0)$ is an equilibrium point which is hyperbolic since Df acts on the tangent plane with the same linear map, and this map has eigenvalues and eigenvectors

$$\lambda^{u,s} = \frac{3}{2} \pm \frac{5}{2}\sqrt{2} \quad , \quad v^{u,s} = \begin{pmatrix} 1 \\ \frac{1}{2} \pm \frac{1}{2}\sqrt{5} \end{pmatrix}. \tag{1.9}$$

Thus, the map f has expanding (unstable) and contracting (stable) bundles, E^u and E^s, along the span of each eigenvector. Notice that the map f preserves the integer lattice; hence, we may consider f as a map on $\mathbb{R}^2/\mathbb{Z}^2$, *i.e.*, the torus T^2. Since f has determinant 1, the induced map on T^2 is invertible. While the action of f on $\mathbb{R}^2$ is rather bland, its action on T^2 is quite interesting: the stable and unstable directions (E^s and E^u) have irrational slopes, so these project down to invariant manifolds on T^2 which wind about the torus densely: see Figure 1.12. Furthermore, the periodic points of f on T^2 are dense, since any pair of rational numbers with the same denominator gives the coordinates of a periodic point.

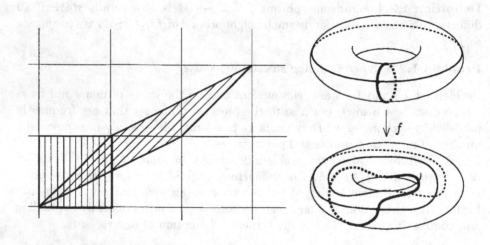

Figure 1.12: The action of the map f on T^2.

Remark 1.2.8 The map of Example 1.2.7 is hyperbolic on all of T^2, hence it is Anosov. We will return to this *toral Anosov map* in §2.3.4.

Notice in Example 1.2.7 that the stable and unstable bundles in the tangent space are mimicked in the base space by invariant manifolds (the projection of E^s and E^u) on which the map is uniformly contractive or expansive. For a map on M with a hyperbolic structure on some invariant set Λ, the splitting of the tangent bundle TM_Λ into invariant stable and unstable bundles projects down to give invariant stable and unstable manifolds in M. This is the content of one of the key results of this field: the Stable Manifold Theorem.[4]

Theorem 1.2.9 (The Stable Manifold Theorem: Hirsch, Pugh, and Shub [84]) *Given a diffeomorphism $f : M \to M$ with a hyperbolic invariant set Λ, for each $x \in \Lambda$, the sets*

$$
\begin{aligned}
W^s(x) &= \{y \in M : \lim_{n \to \infty} \|f^n(y) - f^n(x)\| = 0\}, \\
W^u(x) &= \{y \in M : \lim_{n \to -\infty} \|f^n(y) - f^n(x)\| = 0\},
\end{aligned}
\tag{1.10}
$$

are smooth, injective immersions of the bundles E_x^s and E_x^u respectively. In addition, $W^s(x)$ and $W^u(x)$ are tangent to the bundles at x: $T(W^s(x))_x = E_x^s$ and $T(W^u(x))_x = E_x^s$. The sets $W^s(x)$ and $W^u(x)$ are known as the stable and unstable manifolds of x.

Remark 1.2.10 The notion of *local* stable and unstable manifolds is also useful. Given f as in Theorem 1.2.9, the local stable and unstable manifolds are defined as:

$$
\begin{aligned}
W^s_{loc}(x) &= \{y \in M : \lim_{n \to \infty} \|f^n(y) - f^n(x)\| = 0 \\
&\qquad \text{and } \|f^n(y) - f^n(x)\| < \epsilon \;\; \forall n \geq 0\}, \\
W^u_{loc}(x) &= \{y \in M : \lim_{n \to -\infty} \|f^n(y) - f^n(x)\| = 0 \\
&\qquad \text{and } \|f^n(y) - f^n(x)\| < \epsilon \;\; \forall n \leq 0\},
\end{aligned}
\tag{1.11}
$$

for ϵ of "appropriately" small size.[5] Theorem 1.2.9 then states that $W^s_{loc}(x)$ and $W^u_{loc}(x)$ are tangent to E_x^s and E_x^u.

Theorem 1.2.9 is a very strong result, which we will rely upon frequently to describe the dynamics on a hyperbolic invariant set. The real issue then is ascertaining the smallest invariant subset of M which contains "all" of the essential dynamics of the flow, and then considering systems in which this piece is hyperbolic. Through work of Smale, Shub, and others [165, 162], we know this essential piece to be the *chain-recurrent set*.

Definition 1.2.11 Given a map $f : M \to M$, a point $x \in M$ is *chain-recurrent* for f if, for any $\epsilon > 0$, there exists a sequence of points $\{x = x_1, x_2, \ldots, x_{n-1}, x_n = x\}$ such that $\|f(x_i) - x_{i+1}\| < \epsilon$ for all $1 \leq i \leq n - 1$. The *chain-recurrent set*, $\mathcal{R}(f)$, is the set of all chain-recurrent points on M.

Remark 1.2.12 The chain-recurrent set $\mathcal{R}(f)$ is closed and invariant.

When one has a hyperbolic chain-recurrent set, there is a sort of prime decomposition theorem for the associated dynamics:

---

[4]The Stable Manifold theorem was proved in stages, by several authors, starting with the cases of Λ a fixed point or periodic orbit. Theorem 1.2.9 is a rather general statement.

[5]There is some ambiguity about the size of ϵ – an appropriate size is usually clear from the context.

Theorem 1.2.13 (Smale [165]) *Given a diffeomorphism $f : M \to M$ having a hyperbolic chain-recurrent set, $\mathcal{R}(f)$ is the union of disjoint basic sets, $\mathcal{B}_i$, $i = 1, 2, \ldots, N$. Each $\mathcal{B}_i$ is closed, invariant, and contains a dense orbit. The periodic orbit set of each $\mathcal{B}_i$ is dense within $\mathcal{B}_i$.*

In later chapters, we will often deal with systems which have hyperbolic chain-recurrent sets of various types. One more condition is often required: a map is said to satisfy the *strong transversality condition* if, for all $x, y \in \mathcal{R}(f)$, the stable and unstable manifolds, $W^s(x)$ and $W^u(y)$, are transverse. This condition is important in the definition of *Morse-Smale* and *Smale* diffeomorphisms. A *Smale diffeomorphism* is one which has a zero-dimensional hyperbolic chain-recurrent set satisfying the strong transversality condition, while a *Morse-Smale diffeomorphism* is a Smale diffeomorphism for which the chain-recurrent set is finite.

Working with hyperbolic chain-recurrent sets and transversality has permitted a partial solution of the stability problem (Problem 1.2.5):

Theorem 1.2.14 (Robbin [150], Robinson [151]) *Any diffeomorphism $f : M \to M$ having a hyperbolic chain-recurrent set and satisfying the transversality condition, is structurally stable.*

Continuous dynamics

A map can be considered as a $\mathbb{Z}$-action on M. A continuous analogue to a map is an $\mathbb{R}$-action, or a *flow*.

Definition 1.2.15 A *flow* on a manifold M is a continuous map $\Phi : \mathbb{R} \times M \to M$ satisfying the following conditions:

1. $\phi_t \equiv \Phi(t, -) : M \to M$ is a homeomorphism of M for all t;

2. $\phi_0 = \mathrm{id}_M$, that is, $\phi_0(x) = x$ for all $x \in M$;

3. $\phi_t(\phi_s(x)) = \phi_{t+s}(x)$ for all $s, t \in \mathbb{R}$.

While flows and maps are fundamentally different objects, in certain instances they can be related. Given a map $f : M \to M$, one can define the *suspension flow* of f to be the quotient space of $M \times \mathbb{R}$ with the trivial flow $\phi_t(x, s) = (x, s + t)$ via identifying (x, s) with $(f(x), s + 1)$. The flow ϕ_t passes to a *suspension flow*, $\tilde{\phi}_t$, acting on the *mapping torus* $\tilde{M} = M \times \mathbb{R}/(x, s) \sim (f(x), s + 1)$. In the case where f is isotopic to the identity map, $\tilde{M}$ is homeomorphic to $M \times S^1$, hence the name.

Conversely, given a flow ψ_t on a closed manifold S, we say that S has a *local cross section* (or *Poincaré section*) if there exists a closed codimension-one submanifold $\Pi \subset S$ which transversely intersects the flow at every point of Π. In the case where some subset $U \subset \Pi$ consists of orbits which return to Π in finite time, there is a well-defined *return map* (or *Poincaré map*) $r : U \to \Pi$ which assigns to a point $p \in U$ the image $\psi_{T(p)}(p)$, where $T(p)$, the *return time*, is the smallest $t > 0$ such that $\psi_t(p) \in \Pi$. In the case where Π intersects all flow lines of ϕ_t, we say that Π is a *global cross section*. Clearly, taking the (appropriate) Poincaré section is the inverse of suspending a map. The study of iterated mappings assumed its central importance in dynamics after Poincaré developed the technique of cross-sections and return maps to study periodic orbits in flows generated by ordinary differential equations: examples appear throughout the remainer of this text, most notably in Chapter 4.

When passing to flows, many of the definitions of §1.2.1 carry over with the obvious modifications: *e.g.*, invariant sets, periodic orbits, etc. A few definitions require additional explanation:

Definition 1.2.16 An invariant set Λ for a flow ϕ_t on M is *hyperbolic* if there exists a continuous ϕ_t-invariant splitting of the tangent bundle TM_Λ into $E_\Lambda^s \oplus E_\Lambda^u \oplus E_\Lambda^c$ with

$$\|D\phi_t(v)\| \leq Ce^{-\lambda t}\|v\| \quad \forall\, v \in E_\Lambda^s, \quad \forall t > 0,$$
$$\|D\phi_{-t}(v)\| \leq Ce^{-\lambda t}\|v\| \quad \forall\, v \in E_\Lambda^u, \quad \forall t > 0, \qquad (1.12)$$
$$\left.\frac{d\phi_t}{dt}\right|_{t=0}(x) \text{ spans } E_x^c \quad \forall x \in \Lambda,$$

for some fixed $C > 0, \lambda > 1$. The one-dimensional "center" direction E_x^c is tangent to the orbit itself at each point.

Definition 1.2.17 Let $X \subset \Lambda$ be a subset of a hyperbolic invariant set of a flow ϕ_t on M. Then the *stable* and *unstable manifolds* of X in M are given by

$$\begin{aligned} W^s(X) &= \{y \in M : \lim_{t\to\infty} \|\phi_t(X) - \phi_t(y)\| = 0\}, \\ W^u(X) &= \{y \in M : \lim_{t\to-\infty} \|\phi_t(X) - \phi_t(y)\| = 0\}. \end{aligned} \qquad (1.13)$$

The *local* stable and unstable manifolds of a set X are given by:

$$\begin{aligned} W_{loc}^s(X) &= \{y \in M : \lim_{t\to\infty} \|\phi_t(y) - \phi_t(X)\| = 0 \\ &\qquad \text{and } \|\phi_t(y) - \phi_t(X)\| < \epsilon \;\forall t \geq 0\}, \\ W_{loc}^u(X) &= \{y \in M : \lim_{t\to-\infty} \|\phi_t(y) - \phi_t(X)\| = 0 \\ &\qquad \text{and } \|\phi_t(y) - \phi_t(X)\| < \epsilon \;\forall t \leq 0\}, \end{aligned} \qquad (1.14)$$

For ϵ an "appropriately" small positive number.

Remark 1.2.18 Given γ a periodic orbit for a flow ϕ_t, the local stable and unstable manifolds can carry additional information. Consider the case where, say, $W_{loc}^s(\gamma)$ has dimension two: then, the local stable manifold is a *ribbon* containing γ as a core. This ribbon is homeomorphic to either an annulus or a Möbius band, yielding an *untwisted* or *twisted* periodic orbit respectively. We use such information in §3.1, §4.1, and §5.3.

Definition 1.2.19 Given a flow ϕ_t on M, a point $x \in M$ is *chain-recurrent* for ϕ if, for any $\epsilon > 0$, there exists a sequence of points $\{x = x_1, x_2, \ldots, x_{n-1}, x_n = x\}$ and real numbers $\{t_1, t_2, \ldots, t_{n-1}\}$ such that $t_i > 1$ and $\|\phi_{t_i}(x_i) - x_{i+1}\| < \epsilon$ for all $1 \leq i \leq n - 1$. The *chain-recurrent set*, $\mathcal{R}(\phi)$, is the set of all chain-recurrent points on M.

The Stable Manifold Theorem for flows is entirely analogous to Theorem 1.2.9, and Theorem 1.2.13 holds as stated for flows with hyperbolic chain-recurrent sets. The definitions of Morse-Smale and Smale flows follows with one modification: their chain-recurrent sets are *one*-dimensional, since these are flows. Hence, a Morse-Smale flow is a flow which has a finite number of hyperbolic fixed points and periodic orbits, all of whose stable and unstable manifolds intersect transversally: see Appendix A.

1.2.2   Symbolic dynamics

One of the most remarkable – and fortunate – properties of complicated hyperbolic invariant sets is the description they admit via *symbolic dynamics*. This theory has a long history, beginning with its use by Hadamard in describing closed geodesics [80], and continuing in the work of Morse [133, 134].

Shifts and subshifts

Let $\boldsymbol{\alpha} = \{x_1, x_2, \ldots, x_N\}$ be an *alphabet* of N *letters*. Denote by Σ_N the space of bi-infinite symbol sequences in $\boldsymbol{\alpha}$:

$$\Sigma_N = \{\ldots a_{-2}a_{-1}.a_0a_1a_2\ldots : a_i \in \boldsymbol{\alpha}, \ \forall i \in \mathbb{Z}\} = \boldsymbol{\alpha}^{\mathbb{Z}}. \qquad (1.15)$$

Points in Σ_N will be called *itineraries*. The space Σ_N is given the product topology and can be endowed with a metric as follows. If $\mathbf{a} = (a_i)_{i=-\infty}^{\infty}$ and $\mathbf{b} = (b_i)_{i=-\infty}^{\infty}$ are itineraries, then the distance $d(\mathbf{a}, \mathbf{b})$ is

$$d(\mathbf{a}, \mathbf{b}) = \sum_{n=-\infty}^{\infty} \frac{\delta(n)}{2^{|n|}}, \quad \text{where} \quad \delta(n) = \left\{ \begin{array}{ll} 0 & : a_n = b_n \\ 1 & : a_n \neq b_n \end{array} \right. . \qquad (1.16)$$

Under this metric, points in Σ_N are close when their symbol sequences agree on large blocks forwards and backwards from the "midpoint" a_0.

Define the *shift map* $\sigma : \Sigma_N \to \Sigma_N$ as follows:

$$\sigma(\ldots a_{-2}a_{-1}.a_0a_1a_2\ldots) = \ldots a_{-1}a_0.a_1a_2a_3\ldots. \qquad (1.17)$$

Under the product topology, the shift map σ is a homeomorphism. The dynamical system (Σ_N, σ) is called the *full N-shift*.

Given $\mathcal{A}$ an N by N matrix of zeros and ones, an itinerary $\mathbf{a} = \ldots a_{-1}.a_0a_1\ldots$ is *admissible* with respect to $\mathcal{A}$ at i if, for $a_ia_{i+1} = x_jx_k$ (where $j, k \in \{1, 2, \ldots, N\}$), $\mathcal{A}(j, k) = 1$. Any itinerary $\mathbf{a}$ which is admissible with respect to $\mathcal{A}$ at all i is called *admissible*.

Definition 1.2.20 Given $\mathcal{A}$ an N by N matrix in zeros and ones, the *subshift of finite type* associated with $\mathcal{A}$ is the dynamical system $(\Sigma_{\mathcal{A}}, \sigma)$, where $\Sigma_{\mathcal{A}} \subset \Sigma_N$ is the set of admissible itineraries and σ is the shift map. The matrix $\mathcal{A}$ is known as the *transition matrix* for $\Sigma_{\mathcal{A}}$, since it specifies those transitions between symbols that are possible within a sequence.

Example 1.2.21 Consider the subshift of finite type associated with the transition matrix

$$A = \begin{bmatrix} 1 & 1 & 0 \\ 1 & 1 & 1 \\ 1 & 1 & 1 \end{bmatrix}. \tag{1.18}$$

Then the system (Σ_A, σ) consists of all bi-infinite sequences in $\{x_1, x_2, x_3\}$ not containing $x_1 x_3$ as a subword.

Remark 1.2.22 An alternative to Definition 1.2.20 comes from graph theory. Let Γ be a directed (oriented) finite graph with vertex set $v = \{v_i\}$ and edge set $e = \{e_j\}$, such that there exists at most one edge connecting any ordered vertex pair in $v \times v$. Then the space of bi-infinite, continuous, directed paths in Γ can be put in bijective correspondence with all bi-infinite symbol sequences in $\{v_i\}$ admissible with respect to a transition matrix A_v, where $A_v(i,j) = 1$ if and only if there is a continuous path from v_i to v_j. Alternatively, directed paths in Γ can also be represented by symbol sequences in the edge labels $\{e_j\}$, where the transition matrix A_e satisfies $A_e(i,j) = 1$ if and only if there the tip of the edge e_i meets the tail of the edge e_j. In general, these matrices, A_v and A_e, will differ. Thus, since the space of paths on Γ is the same, we have shown the existence of different subshifts which are nevertheless conjugate: see Figure 1.13.

Figure 1.13: The vertex graph (left) and the edge graph (right) associated to the 2×2 matrix A, where $A(i,j) = 1$ for all i, j.

Symbolic dynamics and subshifts of finite type are very concrete — one can combinatorially determine all the periodic orbits, fixed points, etc. symbolically. On the other hand, given any bi-infinite "random" sequence of ones and twos, there is an orbit in the full 2-shift whose dynamics precisely follows this sequence of x_1's and x_2's; hence, these systems can encode complicated dynamics.

Our interest in symbolic dynamics lies in the fact that they capture the dynamics of hyperbolic invariant sets of maps.

Theorem 1.2.23 (Bowen [26]) *Let $f : M \to M$ be a diffeomorphism with a hyperbolic chain-recurrent set $\mathcal{R}$ and $\Lambda \subset \mathcal{R}$ a basic set. Then, there exists a semiconjugacy $h : \Sigma_A \to \Lambda$ between Λ and a subshift of finite type. That is, h is a continuous surjection with $h\sigma = fh$. If Λ is zero-dimensional then h is a homeomorphism; i.e., h is a conjugacy.*

For details of the proof of Theorem 1.2.23, see Bowen's work [26], or the reformulations in [53, 162]. The essential tools for Theorem 1.2.23 are *rectangles* and *Markov partitions*, both objects which will be of great use to us in Chapter 2.

Definition 1.2.24 For f a diffeomorphism and Λ a hyperbolic basic set, a closed (not-necessarily connected) set $R \subset \Lambda$ is a *rectangle* provided:

1. $W^s_{loc}(x) \cap W^u_{loc}(y) \in R$ is a single point for all $x, y \in R$; and

2. $\text{int}(R)$ is dense in R.

Definition 1.2.25 Let f be a diffeomorphism, Λ a hyperbolic basic set for f, and Ω a finite collection of rectangles R_i. Let $W^s(x, R_i) \equiv W^s_{loc}(x) \cap R_i$ and $W^u(x, R_i) \equiv W^u_{loc}(x) \cap R_i$. Then Ω is a *Markov partition* for f if:

1. $\Lambda = \cup_i R_i$;

2. $\text{int}(R_i) \cap \text{int}(R_j) = \emptyset$;

3. for $x \in \text{int}(R_i)$ and $f(x) \in \text{int}(R_j)$,

$$f(W^s(x, R_i)) \subset W^s(f(x), R_j)), \qquad W^u(f(x), R_j) \subset f(W^u(x, R_i)),$$

4. for $x \in \text{int}(R_i) \cap f^{-1}(\text{int}(R_j))$,

$$\text{int}(R_j) \cap f\left[W^u(x, \text{int}(R_i))\right] = W^u(f(x), \text{int}(R_j)),$$
$$\text{int}(R_i) \cap f\left[\text{int}(W^s(f(x), \text{int}(R_j)))\right] = W^s(x, \text{int}(R_i)).$$

Condition 4 is excluded in many definitions; however, any partition satisfying the first three can be refined to have rectangles of arbitrarily small diameter, implying Condition 4 [153, Lemma 6.8].

Remark 1.2.26 Although rectangles are not necessarily connected, or even locally connected, they can *usually* be thought of as disjoint rectangular simplices: see Example 1.2.28 below and the proof of Lemma 2.2.5. A Markov partition gives rise to a subshift in the following manner: let $\{R_i\}_1^N$ be a Markov partition for a basic set Λ of f as above. Define the $N \times N$ matrix $\mathcal{A}$ by

$$\mathcal{A}(i,j) = \begin{cases} 1 : & f(R_i) \cap R_j \neq \emptyset \\ 0 : & f(R_i) \cap R_j = \emptyset \end{cases}. \tag{1.19}$$

Then, the content of Theorem 1.2.23 is that the subshift of finite type $(\Sigma_\mathcal{A}, \sigma)$ is semiconjugate to (Λ, f), and conjugate in the case when Λ is zero-dimensional.

Remark 1.2.27 There exists an analogue of Theorem 1.2.23 for non-invertible maps. Let $\Sigma_\mathcal{A}^+$ denote the space of *semi-infinite* symbol sequences admissible with respect to $\mathcal{A}$. If we redefine the shift map as $\sigma : (a_0 a_1 a_2 \ldots) \mapsto (a_1 a_2 \ldots)$, then the system $(\Sigma_\mathcal{A}^+, \sigma)$ is a *one-sided subshift of finite type*. The analogue to Theorem 1.2.23 then holds for hyperbolic noninvertible maps and one-sided subshifts.

Example 1.2.28 (Smale's horseshoe) Consider a map $f : I \times I \to \mathbb{R}^2$ on the square given as in Figure 1.14. The map acts linearly on the horizontal strips labeled H_1 and H_2, stretching by a factor $\lambda^u > 2$ in the vertical direction and compressing by $\lambda^s < \frac{1}{2}$ in the horizontal direction, while bending the entire square into a "horseshoe."

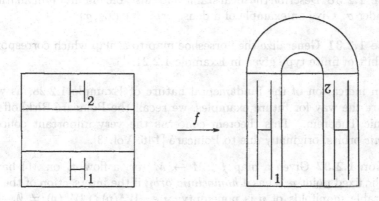

Figure 1.14: The Smale horseshoe map.

Let Λ denote the set of points in $I \times I$ which remain in $I \times I$ under all forwards and backwards iterates of f. This set is invariant and is contained in $H_1 \cup H_2$. Because of the linear action on horizontal strips, the local stable manifold of a point $x \in \Lambda$ is a horizontal line segment passing through x. Similarly, the local unstable manifold of x is a vertical line segment through x. It follows that Λ is a closed hyperbolic invariant set for f.

It is left as an exercise for the reader to show that the intersection of Λ with the (literal) rectangles H_1 and H_2 provides a Markov partition for $f|_\Lambda$. Since Λ is the cartesian product of two Cantor sets in the interval, it follows that Λ is zero-dimensional and, via Theorem 1.2.23, has dynamics conjugate to the subshift of finite type induced by the Markov partition: in this case, the full 2-shift. By writing down bi-infinite sequences of symbols, we can immediately conclude that there are, *e.g.*, two fixed points, a countable infinity of periodic orbits, an uncountable number of nonperiodic orbits, and an orbit of f dense in Λ.

Example 1.2.28 is fundamental to the study of complicated dynamics, since it is perhaps the simplest example of a nontrivial hyperbolic set. Moreover, it occurs widely in dynamical systems modeling physically relevant processes, including Poincaré maps for periodically forced oscillators (*cf.* [76] and §2.3.2 below). In subsequent chapters, we will consider the suspension of the horseshoe map f and regard the periodic orbits as knots. Symbolic dynamics will then give us a language for describing these knots. To the readers unfamiliar with the horseshoe, we suggest that either (1) they consult a good reference for more information (*e.g.*[153, 41, 76]); and/or (2) they complete the following exercises to strengthen understanding of this important example:

Exercise 1.2.29 Draw a picture of the second iterate of f, as well as its inverse; then, prove that Λ is zero-dimensional.

Exercise 1.2.30 Describe the local stable and unstable manifolds of an itinerary in Σ_2 under σ. Give an example of a dense orbit for (Σ_2, σ).

Exercise 1.2.31 Generalize the horseshoe map to a map which corresponds to the subshift of finite type given in Example 1.2.21.

As an indication of the fundamental nature of Example 1.2.28, as well as to prepare the way for future examples, we recall the Poincaré-Birkhoff-Smale homoclinic Theorem. This theorem concerns the very important concept of *homoclinic* orbits, originally due to Poincaré [146, Vol. 3].

Definition 1.2.32 Given a map $f : M \to M$ (or, a flow ϕ_t on M) having a hyperbolic fixed point p, p has a *homoclinic orbit* if the intersection of the stable and unstable manifolds of p is nonempty: *i.e.*, $W^s(p) \cap W^u(p) \neq \emptyset$. In the case of a map, we distinguish between *transverse* homoclinic orbits, for which $T_x W^u(p) \oplus T_x W^s(p) = T_x M$ for all $x \in W^s(p) \cap W^u(p)$, and *nontransverse* homoclinic orbits, for which this condition fails.

Theorem 1.2.33 (The Poincaré-Birkhoff-Smale Homoclinic Theorem [146, 18, 164]) *Let $f : \mathbb{R}^2 \to \mathbb{R}^2$ be a diffeomorphism with p a fixed point supporting a transverse homoclinic orbit. Then, for some $N > 0$, f^N contains a Smale horseshoe in a neighborhood of the homoclinic orbit.*

Remark 1.2.34 By "containing a horseshoe" we mean that there exists a compact invariant subset near the homoclinic orbit which is conjugate to the map of Example 1.2.28. Hence, from very general hypotheses one can apply symbolic dynamics to describe and understand complicated dynamics. This perspective will be of use in the remainder of this book as we seek to describe and understand knotted periodic orbits in flows.

Topological entropy

The question arises which shifts or subshifts are equivalent up to conjugacy (*cf.* Remark 1.2.22). While this problem was completely solved by Williams [191], an earlier result gave rise to an easily computable invariant known as *topological entropy*. The original definition of topological entropy for a map f acting on a compact manifold M considered the growth rates of open covers of M under the action of f. We will use an alternate definition due to Bowen [26].

Definition 1.2.35 Given $f : M \to M$ a diffeomorphism with compact invariant set Λ, an integer $n > 0$, and a real number $\epsilon > 0$, an (n, ϵ)-*separated set* $S \subset \Lambda$ is a set for which any two distinct points x and y in S satisfy $d(f^k(x), f^k(y)) > \epsilon$

for some $0 \leq k < n$. Define $s(n, \epsilon)$ to be the maximum cardinality of any (n, ϵ)-separated subset of Λ. Then, the *topological entropy* of f on Λ is given as

$$h(f) = \lim_{\epsilon \to 0} \lim_{n \to \infty} \sup \frac{\log s(n, \epsilon)}{n}. \qquad (1.20)$$

Definition 1.2.35 is by no means transparent. An (n, ϵ)-separated set is a collection of points which avoid one another (up to ϵ) within the initial segment of the orbit (up to n iterates). On a compact manifold M, every such set must be finite. The entropy is thus the limit of the growth rate (in n) of the maximal number of orbits which separate, as we increase our sensitivity to separation ($\epsilon \to 0$).

Part of the difficulty in understanding Definition 1.2.35 is in ascertaining what topological entropy measures. In short, a map with positive entropy has a great deal of "activity" — the number of orbits which are separated under the action of f grows at an exponential rate. This implies that both stretching (for separation) and folding (for compactness) actions are necessary for complicated dynamics, *cf.* Example 1.2.28. Alternatively, a map which has zero entropy (*e.g.*, an isometry) would indicate a relatively small degree of complicated dynamics. A rough generalization is that positive topological entropy signals "chaotic" dynamics.

Remark 1.2.36 Two maps on compact spaces which are conjugate must have the same entropy, since the conjugacy is a uniformly continuous homeomorphism which preserves $s(n, \epsilon)$ after a change of scale in ϵ. Hence, topological entropy is a dynamical invariant. Topological entropy for flows is less well-defined: if we define the entropy of a flow to be the entropy of the time-one map, then we can at least distinguish zero-entropy from positive-entropy flows.

Calculating entropy is in general a difficult task: fortunately, the entropy of the shifts and subshifts of §1.2.2 are readily computed.

Theorem 1.2.37 *Let Σ_A denote the subshift of finite type associated with the matrix A. Then the entropy of the shift map σ is the log of the spectral radius of A.*

Theorem 1.2.37 relies upon the Perron-Frobenius Theorem for matrices with positive entries [143, 60]. A nice proof of Theorem 1.2.37 can be found in [153].

Example 1.2.38 The entropy of the full 2-shift is $\log(2)$, since the full 2-shift has as transition matrix a 2×2 matrix with ones in each entry. Thus by Remark 1.2.36 we know that the Smale horseshoe map has entropy equal to $\log(2)$.

In the Appendix, we will use entropy to characterize knots and links, partitioning the set of links into zero-entropy and positive-entropy links.

1.2.3   Bifurcations and one-dimensional maps

We have thus far considered the case in which the dynamical system (or its chain-recurrent set) is hyperbolic. Now suppose we have a family of systems dependent upon a parameter $\mu \in \mathbb{R}^n$. By Theorem 1.2.14, as long as the system is hyperbolic, varying the parameter has no qualitative effect. However, if we specify merely that the system have the appropriate hyperbolic structure for a certain μ_0, then varying the parameter μ may alter it drastically — fixed points, periodic orbits, and basic sets may appear or vanish in *bifurcations*.

We review the simplest types of bifurcations in order to provide a language with which to describe the creation of knotted orbits in parametrized families of three-dimensional flows in Chapter 4. For more complete expositions, see [39, 76]. The following three examples represent the simplest types of bifurcations which can be embedded in one-parameter families of one-dimensional maps:

Example 1.2.39 (saddle-node bifurcation) Let $f_\mu : \mathbb{R}^1 \to \mathbb{R}^1$ be an otherwise generic map whose derivative satisfies $f_0'(0) = 1$: *e.g.*, $x \mapsto x + (\mu - x^2)$. Then the bifurcation at $\mu = 0$, in which two stable equilibria are created, is called a *saddle node bifurcation*. For $\mu < 0$ there are no fixed points for f. As μ increases through zero, a pair of hyperbolic fixed points of opposite stability branches out from the origin.

Example 1.2.40 (pitchfork bifurcation) Although the saddle-node bifurcation is the generic one-parameter bifurcation for $f_0'(0) = 1$, other bifurcations are possible under specific restrictions on the class of maps considered. For instance, assume that $f : \mathbb{R}^1 \to \mathbb{R}^1$ is generic in the class of maps which is invariant under the symmetry transformation $x \mapsto -x$: *e.g.*, $x \mapsto x + (\mu x - x^3)$. Then, by symmetry, the origin must be a fixed point for all μ. In this case, there is a *pitchfork bifurcation* at $\mu = 0$. For $\mu < 0$, the origin is an isolated hyperbolic fixed point. As μ increases through zero, the origin changes stability and simultaneously sheds two fixed points, each acquiring the stability type the origin had for μ negative.

Example 1.2.41 (period-doubling bifurcation) Let $f_\mu : \mathbb{R}^1 \to \mathbb{R}^1$ be a generic map whose derivative satisfies $f_0'(0) = -1$: *e.g.*, $x \mapsto -x - \mu x + x^3$. Then the bifurcation at $\mu = 0$ is called a *period-doubling bifurcation*, since a period two orbit is created. For $\mu < 0$ there is an isolated hyperbolic fixed point at the origin. As μ increases through zero, the origin changes stability and a period two orbit branches away from the origin.

Remark 1.2.42 The three examples above may come in different flavors: for example, the signs of the nonlinear terms may differ. Also, these examples are not confined to bifurcations of one-dimensional maps. Arbitrary maps can exhibit, *e.g.*, a saddle-node bifurcation. This theory involves the construction of one-dimensional *center manifolds*, which capture the bifurcating orbits. See, for example, the introductory texts [153, 76, 9, 34].

Exercise 1.2.43 Given the three bifurcations of fixed points presented above, explain via Poincaré maps what happens to a periodic orbit in a flow which undergoes a saddle-node, pitchfork, or period-doubling bifurcation. Then, re-consider the statement at the beginning of this chapter, that a period-doubled orbit in a three-dimensional flow gives rise to a 2-cable of the knot.

Examples 1.2.39 and 1.2.41 are *codimension one* bifurcations: they occur stably for generic one-parameter families of maps. (In the absence of symmetry, the pitchfork bifurcation of Example 1.2.40 is of codimension two, since two conditions, one on the eigenvalue and one on the quadratic term (that it vanishes), must simultaneously be met.) There is a third important codimension one bifurcation:

Example 1.2.44 (Hopf bifurcation) The Hopf bifurcation for a periodic orbit involves a complex conjugate pair of eigenvalues for the linearized Poincaré map and thus can occur only for maps *of dimension two or greater*. The truncated normal form, analogous to the one-dimensional versions above, is most naturally expressed in polar coordinates:

$$\begin{pmatrix} r \\ \theta \end{pmatrix} \overset{F_\mu}{\mapsto} \begin{pmatrix} r(1 + \mu - r^2) \\ \theta + \varphi + br^2 \end{pmatrix}, \tag{1.21}$$

the linearized mapping in cartesian form being

$$F_\mu = (1 + \mu) \begin{bmatrix} \cos\varphi & -\sin\varphi \\ \sin\varphi & \cos\varphi \end{bmatrix} : \tag{1.22}$$

a matrix with eigenvalues $\lambda, \overline{\lambda} = (1 + \mu)e^{\pm i\varphi}$, which rotates by the angle φ and dilates by the factor $1 + \mu$. It is easy to check that, for $\mu < 0$, (1.21) has an isolated hyperbolic sink at the origin, from which an attracting invariant circle $r = \sqrt{\mu}$ bifurcates as μ increases through zero. On this circle, points are rigidly rotated through the angle $\varphi + b\mu$. When this quantity is rational (mod 2π) the invariant circle is filled with periodic points; when irrational, with dense, quasi-periodic orbits.

As the orbits created in a Hopf bifurcation lie on the boundary of a tubular neighborhood of the periodic orbit (that is, a torus), any periodic orbits are cables of the original knot: we return to this in Chapter 4.

When working with families of one-dimensional maps, the symbolic theory of subshifts in §1.2.2 can be used effectively to encode sequences of bifurcations as a parameter is varied. To do so, we must specify a coordinate system on symbol sequences induced by the one-dimensional map. These coordinates foreshadow a similar construct to be used for semiflows on branched two-manifolds having one-dimensional return maps. This *kneading theory* will be used in locating periodic orbits and determining their topological properties in later chapters.

To introduce the ideas, consider the two hyperbolic (expanding) maps defined on $I = [0, 1] \subset \mathbb{R}$ of Figure 1.15. In both cases a Markov partition may be based

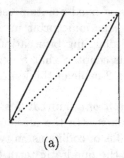
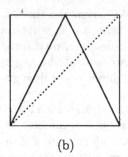

(a)   (b)

Figure 1.15: One dimensional maps: (a) the doubling map, f_D; (b) the tent map, f_T.

on the intervals $I_1 = [0, \frac{1}{2}]$ and $I_2 = [\frac{1}{2}, 1]$, each of which is stretched across the entire interval I by the map. Labeling the intervals x_1 and x_2 and appealing to Theorem 1.2.23, we have a semiconjugacy[6] between f_D (resp. f_T) and a full shift on two symbols, although here it is the *one-sided* shift working on semi-infinite sequences, since one can only iterate the maps forwards (*cf.* Remark 1.2.27).

Under these semiconjugacies, a point $p \in I$ belonging to a periodic orbit of either map corresponds to a sequence formed of repeats of a finite word $\mathbf{a}_p = (a_0 a_1 \ldots a_{K-1})$, of length K equal to the (least) period, in which the symbol a_j takes the value x_1 (resp. x_2) if $f^j(p) \in I_1$ (resp. I_2). The itinerary formed by repeating a word $\mathbf{w}$ will be denoted $\mathbf{w}^\infty$.

To locate points within an orbit, or points of distinct orbits, we introduce the natural "left to right" lexicographical ordering $x_1 \lhd x_2$. Here the two maps reveal a crucial difference. Since f_D is orientation-preserving (both branches have positive slope), simple lexicographical ordering of the itineraries $\mathbf{a}_p^\infty$ and $\mathbf{a}_q^\infty$ will correctly determine the relative positions of the points $p, q \in I$. Essentially here we are comparing binary expansions of p and q, with x_1 and x_2 playing the roles of 0 and 1.

Example 1.2.45 Consider the points $p = \frac{1}{3}$ and $q = \frac{3}{7}$, whose orbits under f_D are $\{\frac{1}{3}, \frac{2}{3}, \frac{1}{3}, \frac{2}{3}, \ldots\}$ and $\{\frac{3}{7}, \frac{6}{7}, \frac{5}{7}, \frac{3}{7}, \frac{6}{7}, \frac{5}{7}, \ldots\}$ respectively. The associated words are: $\mathbf{a}_p^\infty = \{x_1 x_2 x_1 x_2 \ldots\}$ and $\mathbf{a}_q^\infty = \{x_1 x_2 x_2 x_1 x_2 x_2 \ldots\}$. $\mathbf{a}_p^\infty$ and $\mathbf{a}_q^\infty$ first differ at the third symbol, and since $x_1 \lhd x_2$, we see that $\mathbf{a}_p^\infty \lhd \mathbf{a}_q^\infty$, as required.

Turning to the map f_T, we note that orientation is *reversed* for points in I_2. To cope with this, we compare not simple itineraries, but *invariant coordinates*, defined as $\theta(\mathbf{a}) = \theta_1 \theta_2 \ldots \theta_n \ldots$, where $\theta_i = a_i$ if the x_2-parity of $a_1 a_2 \ldots a_{i-1}$

---

[6]Here, the map is a semiconjugacy because points on the boundary $I_1 \cap I_2 = \frac{1}{2}$ admit two distinct symbol sequences $x_2 (x_1)^\infty$ and $x_1 (x_2)^\infty$ (*cf.* the ambiguity in decimal representation of reals). The maps from f_D or f_T to the full shift are conjugacies when restricted to the periodic orbit set. One can also get semiconjugacies if the slope of the map is of absolute value less than one: multiple orbits may share the same symbol sequence.

is even, else $\theta_i = \hat{a}_i$, where $\hat{x}_2 = x_1$ and *vice versa*. Thus θ keeps track of how many visits to the orientation-reversing subinterval the orbit has made.

Example 1.2.46 Again take two points, but now belonging to periodic orbits of f_T: $p = \frac{2}{5}$ and $q = \frac{2}{7}$. The associated words are again: $\mathbf{a}_p^\infty = (x_1 x_2)^\infty$ and $\mathbf{a}_q^\infty = (x_1 x_2^2)^\infty$, but the invariant coordinates are:

$$\theta(x_1 x_2)^\infty = (x_1 x_2^2 x_1^2 x_2)^\infty,$$
$$\theta(x_1 x_2^2)^\infty = (x_1 x_2 x_1^2 x_2 x_1)^\infty.$$

We now correctly have $\theta(\mathbf{a}_q^\infty) \lhd \theta(\mathbf{a}_p^\infty)$.

Thus, extending the definition of θ appropriately for general multi-branch maps to count the number of visits to orientation-reversing subintervals, we have:

Proposition 1.2.47 (Milnor and Thurston [125]) *Let p and q be points on I corresponding to words $\mathbf{a}_p^\infty$ and $\mathbf{a}_q^\infty$ respectively. Then $p < q \Leftrightarrow \theta(\mathbf{a}_p^\infty) \lhd \theta(\mathbf{a}_q^\infty)$.*

We have described the theory for the special cases of piecewise linear maps, but it applies equally well to nonlinear maps; in fact one does not even need the slope to exceed 1 everywhere. If the slope does exceed one on each branch (the map is hyperbolic or expansive), and the subintervals I_j are pairwise disjoint, then the semiconjugacy referred to above becomes a conjugacy.

We call a word $\mathbf{a}$ *minimal* if the invariant coordinate of $\mathbf{w}$ is minimal with respect to $\lhd$ in the invariant coordinates of the shift equivalence class, i.e., $\theta(\mathbf{a}) \trianglelefteq \theta(\sigma^i(\mathbf{a}))$, $\forall i$. In the kneading theory of one dimensional maps, the minimal word is also called the *itinerary* of the orbit. We now briefly review some ideas from this area; for details see [39].

That portion of one-dimensional kneading theory with which we will be concerned seeks to order points on the interval with respect to symbol sequences (as in Proposition 1.2.47) and also to explicitly determine bifurcation sequences for *unimodal* maps of the type illustrated in Figure 1.16, the canonical example of which is the quadratic family:

$$f_\mu : x \mapsto \mu - x^2. \tag{1.23}$$

Upon increasing μ, the nonwandering set of f_μ changes from being empty for $\mu < -\frac{1}{4}$, to having a one-dimensional analogue of a hyperbolic horseshoe for $\mu > 2$. This sequence of bifurcations involves numerous period-doubling and saddle-node bifurcations in an order which displays self-similarity: see [41, 198, 199]. Note that, for $\mu = 2$, a homeomorphism on the interval $[-2, 2]$ takes f_μ into f_T [181], *cf.* [76, §5.6].

The range of the map f_μ is determined by the orbit of the critical point c, which essentially determines the dynamics of the map. We assign to each periodic orbit of f_μ a word which allows us to order bifurcations, much as itineraries

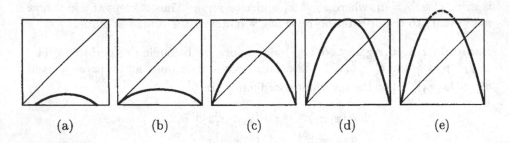

| (a) | (b) | (c) | (d) | (e) |

Figure 1.16: Members of the quadratic family f_μ: (a) $\mu < -\frac{1}{4}$; (b) $\mu = -\frac{1}{4}$; (c) $\mu \in (-\frac{1}{4}, 2)$; (d) $\mu = 2$; (e) $\mu > 2$.

and invariant coordinates permit the ordering of points on the interval I. In-tuitively, this word, the *kneading invariant* $\nu(\mathbf{a})$, is the itinerary of the critical point c for the μ-value at which a point on the orbit $\mathbf{a}$ crosses over c. There are some technicalities regarding which orbits actually contain points that cross c and whether they can be given an associated invariant. These details are unwieldy and largely unnecessary for our purposes: the diligent reader should consult [39, 41, 90].

Given a word $\mathbf{a} = a_1 a_2 \ldots a_n$ (with $n \geq 3$) one can associate such a sequence given by

$$\nu(\mathbf{a}) = \theta(x_1 x_2 (a_1 a_2 \ldots a_{n-2} c x_2)^\infty), \tag{1.24}$$

where $c = x_2$ if the x_2-parity of $a_1 a_2 \ldots a_{n-2}$ is even and $c = x_1$ if it is odd. The two period one orbits have kneading invariant $\nu(x_1) = \nu(x_2) = x_1^\infty$ and the single period two orbit has $\nu(x_1 x_2) = (x_1 x_2)^\infty$.

The important fact concerning kneading invariants and bifurcations of f_μ is:

Proposition 1.2.48 (Milnor and Thurston [125]) *Let $\mathbf{a}_p^\infty$ and $\mathbf{a}_q^\infty$ be the minimal words for periodic orbits of f_μ and let μ_p, μ_q be the μ-values at which these periodic orbits are created. Then, with $\lhd$ as before, $\nu(\mathbf{a}_p) \lhd \nu(\mathbf{a}_q) \Rightarrow \mu_p < \mu_q$.*

Thus, for the quadratic map f_μ, we may completely characterise "which comes first" in the orbit genealogy. As above, the theory works for a more general class of unimodal maps than f_μ, the main requirement being that the maps have negative Schwarzian derivative [163, 39, 76], implying that, for each μ, there is at most one stable periodic orbit.

We have now sketched the requisite background material. In the chapters that follow, we will demonstrate how ideas from knots and links and dynamical systems theory can be drawn together. In doing so, we will be able both to answer questions in dynamical systems and bifurcation theory, and to discover new phenomena in low-dimensional topology.

Chapter 2: Templates

We now proceed with our program to investigate the link of periodic orbits in a three-dimensional flow. In this chapter, we blend the two themes of Chapter 1, the study of knots, and the study of hyperbolic dynamics, to create a tool for analyzing knotted orbits of hyperbolic flows: the *template*. This important tool, whose origins lie within the work of R. F. Williams [192, 193], will be our primary instrument for examining periodic orbit links.

In §2.1 we review the natural role of branched one-manifolds as attractors, foreshadowing the concept of a template. In §2.2, we give a thorough treatment of the Template Theorem of Birman and Williams [24] and then apply this theorem in §2.3 to a variety of important three-dimensional flows. Finally, in §2.4, we construct a set of symbolic tools for describing and manipulating templates and the orbits that they carry.

First, we consider the example which motivated much of this work (*cf.* [193, p. 111]):

Example 2.0.1 Given a three dimensional flow, our main goal is to determine relationships between the link of periodic orbits (as a topological object) and the dynamics and bifurcations of the system. To proceed, we must be able to ascertain which types of knots and links a given flow supports. For a sufficiently complicated flow (*e.g.*, on a basic set of dimension two), there exist a countable infinity of periodic orbits which fill up an attractor densely. In this case, even visualizing the flow may be a challenge.

The following set of ordinary differential equations (ODEs) is known as the *Lorenz system* [114]:

$$\begin{aligned}
\dot{x} &= 10(y - x) \\
\dot{y} &= 28x - y - xz \\
\dot{z} &= -\frac{8}{3}z + xy,
\end{aligned} \qquad (2.1)$$

A numerical integration of the system suggests an *attractor*: all orbits appear to collapse quickly onto a particular subset $\mathcal{L} \subset \mathbb{R}^3$, called the *Lorenz attractor*. The structure of this attractor is unusual: it appears to be two-dimensional, yet is not a manifold. Rather, the attractor $\mathcal{L}$ (illustrated in Figure 2.1) resembles a *branched two-manifold*. Nevertheless, as Lorenz realized at the outset [114], it has infinitely many sheets.

If we wish to understand the periodic orbits of this system, we need only consider those orbits which live on $\mathcal{L}$, since all other orbits appear to converge to $\mathcal{L}$, and hence none of them can be periodic. Thus, heuristically, we can reduce

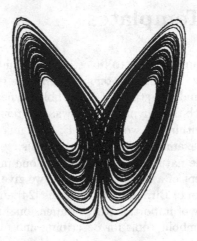

Figure 2.1: The Lorenz attractor (computed via *DsTool*).

the dimension of our problem by one: we need only consider knotted orbits on a branched two-manifold. A *template* is just such a branched two-manifold which "supports" the periodic orbits of a flow. The theory of templates, which we treat in this chapter, is a rigorous method for applying this idea to general hyperbolic flows on three-manifolds.

2.1   Branched manifolds and attractors

In order to motivate the Template Theorem of Section §2.2, we briefly describe the role of branched manifolds as attractors for hyperbolic systems. We begin with a discussion of branched one-manifolds in the dynamics of two-dimensional maps before considering the role of branched two-manifolds, or templates, in the dynamics of three-dimensional flows.

Definition 2.1.1 A *branched one-manifold* is a topological space built locally from a finite number of *branch point charts*, as illustrated in Figure 2.2(a). Each chart has a finite number (≥ 1) of arcs emanating from a *branch point* along both sides of a common tangent.

Example 2.1.2 The branched one-manifold of Figure 2.2(b) is known as the *Plykin branched manifold*, Γ_P.

Branched one-manifolds are a key tool for understanding *expanding attractors* for 2-dimensional maps.

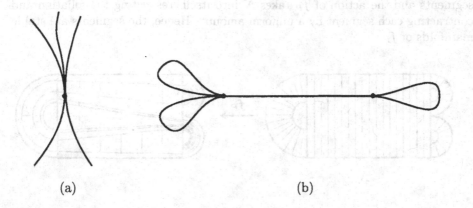

(a)   (b)

Figure 2.2: (a) a branch point chart for a branched one-manifold; (b) the Plykin branched manifold, Γ_P.

Definition 2.1.3 For $f : M \to M$ a diffeomorphism, a set $\Lambda \subset M$ is an *attractor* if there exists a compact set $N \subset M$ such that $\Lambda = \cap_{k=0}^{\infty} f^k(N)$ and Λ is contained in the chain-recurrent set $\mathcal{R}(f)$. If $f|_\Lambda$ has a hyperbolic structure, then Λ is a *hyperbolic attractor*. Finally, Λ is an *expanding attractor* if it is hyperbolic and has topological dimension equal to the dimension of E^u, the unstable bundle.

Williams [192] considered the relationship between expanding attractors and branched manifolds (in any dimension). For two-dimensional maps, the theory boils down to the following:

Theorem 2.1.4 (Williams [192]) *Let $f : M \to M$ be a diffeomorphism on a two-manifold M with $\Lambda \subset M$ an expanding attractor. Then, there exists an embedded branched one-manifold $\Gamma \subset M$ and a noninvertible map $g : \Gamma \to \Gamma$ such that $f|_\Lambda$ is conjugate to the shift map on the inverse limit of (Γ, g).*

Definition 2.1.5 Given a map $g : X \to X$, the *inverse limit*, $\overset{\lim}{\leftarrow} (X, g)$, is given as the space of all bi-infinite sequences $(\ldots, x_{-1}, x_0, x_1, \ldots)$, with $g(x_k) = x_{k+1}$. The shift map associated to $\overset{\lim}{\leftarrow} (X, g)$ takes each x_k to x_{k+1}.

The structure of the expanding attractor Λ in Theorem 2.1.4 is complicated — it is locally the product of $\mathbb{R}^1$ with a Cantor set [192]. However, the map $g : \Gamma \to \Gamma$ is more tractable: *e.g.*, the edges of Γ form a Markov partition for g. To understand the idea behind Theorem 2.1.4, and to provide an analogue for the Template Theorem of §2.2, consider the following:

Example 2.1.6 Construct a map $f_P : \mathbb{R}^2 \to \mathbb{R}^2$ which has the action illustrated in Figure 2.3(a). There is a compact region $N \subset \mathbb{R}^2$ with three holes, each containing a source, and an additional source at "infinity." N is foliated by line segments and the action of f_P takes N into itself, respecting the foliation and contracting each segment by a uniform amount. Hence, the segments are stable manifolds of f_P.

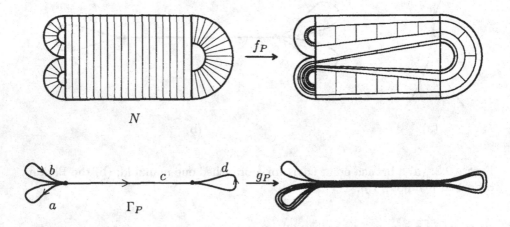

Figure 2.3: (a) The map f_P acting on $N \subset \mathbb{R}^2$ yields the Plykin attractor; (b) The induced map on Γ_P.

The attractor, Λ_P, is given as $\cap_k f_P^k(N)$ and is locally the product of a Cantor set with a one-dimensional local unstable manifold; since Λ_P has topological dimension one (it has empty interior in $\mathbb{R}^2$ yet contains $W^u_{loc}(x)$), it is an expanding attractor. This attractor is called the *Plykin attractor* after [144]. To realize the associated branched one-manifold, collapse each component of $W^s(x) \cap N$ to a point. Since f_P respects the foliation by stable manifolds, the induced map on the branched one-manifold, g_P, is well-defined. It is obvious from Figure 2.3(a) that the branched one-manifold is precisely the Plykin branched one-manifold Γ_P of Example 2.1.2. The dynamics of f_P is captured by the induced map g_P which acts on Γ_P as indicated in Figure 2.3(b).

Exercise 2.1.7 Construct the subshift of finite type associated with the Plykin attractor.

Example 2.1.6 is central to the theme of this chapter: under certain hyperbolicity conditions, Theorem 2.1.4 guarantees that an invariant set for a diffeomorphism on a two-manifold can be "replaced" by a non-invertible map on a branched one-manifold, preserving the essential dynamics. Furthermore, note that, in particular, periodic orbits of the diffeomorphism are treated with respect — they are isotoped along the stable foliation. If we suspend the Plykin map f_P and embed the flow in $\mathbb{R}^3$, periodic orbits become knots and links. The action of collapsing a stable foliation necessarily preserves individual knot and link types.

We will repeat this theme in the next section, substituting a three-dimensional flow for a two-dimensional diffeomorphism, and branched two-manifolds with semiflows for branched one-manifolds with non-invertible maps. We will take the construction a step further in that we do not merely consider "attractors" in three dimensions.

Remark 2.1.8 There is a great deal more to the story of branched manifolds and expanding attractors. In [192], it is shown that an expanding attractor for a diffeomorphism on an $n + 1$-manifold is cojugate to the inverse limit of a diffeomorphism on a *branched n-manifold*, the higher-dimensional analogue of the branched one-manifolds. Several authors have extended or related results in dimensions one (see the literature on *train tracks*) and two (see the work of Christy [37]).

2.2 Templates and the Template Theorem

We now consider an appropriate generalization of the branched one-manifolds of §2.1 for three-dimensional flows, such as that associated with the Lorenz attractor of Example 2.0.1.

Definition 2.2.1 A *template* is a compact branched two-manifold with boundary and smooth expansive semiflow built locally from two types of charts: *joining* and *splitting*. Each chart, as illustrated in Figure 2.4, carries a semiflow, endowing the template with an expanding semiflow, and the gluing maps between charts must respect the semiflow and act linearly on the edges.

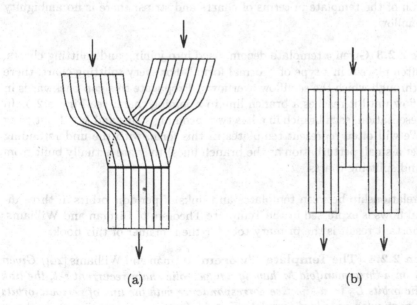

Figure 2.4: (a) a joining chart; (b) a splitting chart.

Definition 2.2.2 Consider a flowbox $I \times I$ having semiflow given by translation in the second coordinate. We define a *joining chart* as the quotient space $((I \times I) \cup (I \times I)) / \{(x,y) = (x,y) : y \leq \frac{1}{2}\}$ with the associated semiflow. Similarly, a *splitting chart* is defined as $I \times I$ minus the set $\{(x,y) : x \in (\frac{1}{3}, \frac{2}{3}), y \in [0, \frac{1}{2})\}$.

The joining chart of Figure 2.4(a) contains two *incoming strips* and one *outgoing strip*, all of which meet tangentially at the *branch line*. The splitting chart of Figure 2.4(b) turns one incoming strip into two outgoing strips as pictured. One builds a template by connecting the free ends of the outgoing strips to the free ends of the incoming strips between charts in a manner to be specified. Since the template must be compact, there may be no "free" ends, and the total number of charts and strips in a template must be finite.

Each chart has an inherited semiflow, by which we mean an irreversible flow (an action of $\mathbb{R}^+$) — a true flow is impossible since reversing the flow just below the branch line would violate uniqueness. The semiflow is *overflowing* in the sense that on the splitting charts, there is a *gap* in the strip through which the semiflow "spills over." Since we are concerned with periodic orbits of the semiflow (i.e., knots), we ignore orbits exiting the template.

We also require that each gluing map connecting the free edge of an outgoing strip to that of an incoming strip be linear. The semiflow as constructed is thus expansive in the sense that the noninvertible one-dimensional return maps for the semiflow induced by the branch lines are expansive maps (these return maps are also piecewise linear and hence uniformly hyperbolic). This being the case, the dynamics (up to conjugacy) are determined uniquely by the combinatorial description of the template in terms of charts and strips: there is no ambiguity in the semiflow.

Remark 2.2.3 Given a template decomposed into joining and splitting charts, we will often place it in a type of "normal form." For every splitting chart, there is a gap through which the semiflow overflows. Propagate this gap backwards in the semiflow until it reaches a branch line in a joining chart: see Figure 2.5. In this representation, each branch line has two incoming strips and $k \geq 1$ outgoing strips. We will often represent templates in this form, with the understanding that (after a small perturbation at the branch lines) they are actually built from joining and splitting charts.

The relationship between templates and links of periodic orbits in three dimensional flows is expressed in the Template Theorem of Birman and Williams. This important result is the primary tool for the remainer of this book.

Theorem 2.2.4 (The Template Theorem: Birman and Williams [24]) *Given a flow ϕ_t on a three-manifold M having a hyperbolic chain-recurrent set, the link of periodic orbits L_ϕ is in bijective correspondence with the link of periodic orbits $L_\mathcal{T}$ on a particular embedded template $\mathcal{T} \subset M$ (with $L_\mathcal{T}$ containing at most two extraneous orbits). On any finite sublink, this correspondence is via ambient isotopy.*

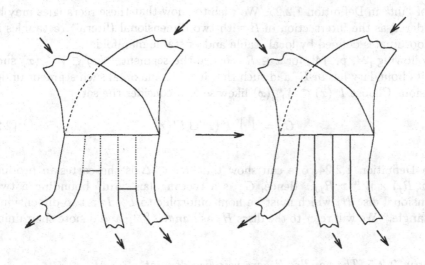

Figure 2.5: By propagating gaps backwards, one obtains a normal form for a template.

Although a proof of Theorem 2.2.4 appears in [24], we include a proof for completeness, as the methods will be of use later.

Proof: Let $\mathcal{R}$ denote the chain-recurrent set of the flow ϕ_t on M. By Theorem 1.2.13, $\mathcal{R}$ decomposes into a finite number of basic sets $\mathcal{B}_i$. The proof depends upon the dimension of each basic set $\mathcal{B}$. Of course, if $\dim(\mathcal{B}) = 0$, there are no periodic orbits and the result is trivially true. We treat the cases $\dim(\mathcal{B}) = 1$ and $\dim(\mathcal{B}) > 1$ in the following subsections:

2.2.1 Case 1: a Markov flowbox neighborhood

Assume that $\dim(\mathcal{B}) = 1$. If we could construct a Poincaré section to the flow on $\mathcal{B}$, then Bowen's theorem on subshifts of finite type (Theorem 1.2.23) would imply that $\mathcal{B}$ is conjugate to a suspended subshift of finite type. Bowen [25] and Bowen and Walters [28] have considered this situation, and have shown that such a cross-section does exist, and can be taken to be a finite union of disjoint discs, $\{\Delta_i\}_{i=1}^N$.

Our strategy (first used in [24]) is to use the properties of rectangles (Definition 1.2.24) and Markov partitions (Definition 1.2.25) to construct a special neighborhood of $\mathcal{B}$ in M.

Step 1: rectangular rectangles

Let $\Delta \equiv \cup_i \Delta_i$ be a collection of embedded discs in M which forms the aforementioned cross-section to $\mathcal{B}$. By Theorem 1.2.23, $\Delta \cap \mathcal{B}$ is a Cantor set with a Markov partition. Let $\Omega \equiv \cup_j R_j$ be the rectangles of the Markov partition (see Definition 1.2.24), and let $\tau : \Omega \to \Omega$ be the Poincaré return map (a homeomorphism). *Note*: since $\Delta \cap \mathcal{B}$ is a Cantor set, one may effectively ignore the rôle of "int" in Definition 1.2.25. We wish to show that these rectangles may be considered as the intersection of $\mathcal{B}$ with two-dimensional (literal) rectangles in the coordinates defined by local stable and unstable manifolds.

Following [24, p. 14], for $x \in R_j$ choose the segment $I^s(x) \subset W^s_{loc}(x)$ such that its boundary lies in R_j and such that it is the maximal such segment under inclusion. Choose $I^u(x) \subset W^u_{loc}(x)$ likewise and consider the set

$$G_j = \bigcup_{x \in R_j} I^s(x) \cup I^u(x). \tag{2.2}$$

From Definition 1.2.24, one can show that $R_j \subset \Delta$ is the cartesian product $W^s(x, R_j) \times W^u(x, R_j)$. Hence, G_j is a rectangular "grid" bounding a two-dimensional disc H_j which must be homeomorphic to $I \times I$: a two-dimensional "rectangle." We will refer to the discs H_j as *handles* [53], and denote their union H.

Lemma 2.2.5 *The handles H_j are pairwise disjoint.*

Proof: Since we may refine the Markov partition Ω to have rectangles of arbitrarily small diameter (see Definition 1.2.25), it remains to show that the rectangles R_i are separated (as sets) by a nonzero distance. However, since the zero-dimensional sets R_i have no boundary in Ω, every $x \in R_i$ is in its interior, and must be bounded away from any other R_j by Condition 2 of Definition 1.2.25 and the fact that rectangles are closed.   □

Step 2: the action of τ on the handles

Extend the return map τ to the handles H. Although not well-defined everywhere, τ is still a homeomorphism on a neighborhood of $\Omega \subset H$.

Lemma 2.2.6 *If $\tau(H_i) \cap H_j \neq \emptyset$, then $\tau(H_i)$ stretches completely across H_j in the unstable direction, and $\tau^{-1}(H_j)$ stretches completely across H_i in the stable direction. Furthermore, $\tau(H_i) \cap H_j$ has at most one connected component.*

Proof: By Condition 3 of Definition 1.2.25, $\tau(W^u(x, R_i)) \supset W^u(\tau(x), R_j)$ for $x \in R_i$. Reverse the flow direction to show the analogous result for stable manifolds. Finally, assume that $\tau(H_i) \cap H_j$ has two components. Then, for $x \in H_i$, $\tau(I^u(x)) \not\subset I^u(\tau(x))$, in violation of Condition 4 of Definition 1.2.25.   □

Let $\mathcal{A}$ be the square matrix with each entry $A(i, j)$ equal to the geometric intersection number of $\tau(H_i)$ with H_j. By Lemma 2.2.6, this number is either zero or one, and $\mathcal{A}$ is the transition matrix for the Markov partition Ω.

Step 3: a Markov flowbox neighborhood

By flowing the handles H_i forwards and backwards in time, we construct a flowbox neighborhood $N(B)$ for the handle set which appears as in Figure 2.6(a): there are a finite number of *incoming* and *outgoing* flowboxes near each H_i.

Consider the transition matrix $\mathcal{A}$: the ith row of $\mathcal{A}$ records which handles H_i flows to. Thus, there are $\sum_j \mathcal{A}(i, j)$ components of $\tau^{-1}(H) \cap H_i$. By Lemma

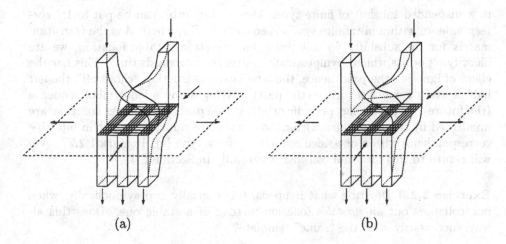

(a)   (b)

Figure 2.6: A Markov flowbox neighborhood of the zero-dimensional basic set.

2.2.6, each of these components stretches completely across H_i in the stable direction. Hence, there are $\sum_j \mathcal{A}(i,j)$ outgoing flowboxes connected to H_i. By reversing the time direction and applying the same argument, one shows that there are $\sum_j \mathcal{A}(j,i)$ incoming flowboxes connected to H_i and stretching in the unstable direction. Since τ is a homeomorphism on Ω and Ω intersects the boundary of each handle H_i, the flow boxes must "line-up" along the edges as in Figure 2.6(a).

Finally, we enlarge the flowbox neighborhood $N(B)$ slightly to have the form of Figure 2.6(b): a small perturbation is all that is required. This is done to fit the joining and splitting chart requirements in Definition 2.2.1.

Lemma 2.2.7 *The periodic orbits of ϕ are in bijective isotopic correspondence with those in an embedded template $\mathcal{T} \subset M$.*

Proof: Given the Markov flowbox neighborhood of $N(B)$ constructed above, one "crushes" a stable foliation as in Example 2.1.6 to obtain a branched manifold. Specifically, form the quotient space given by identifying all points on $W^s(x) \cap N(B)$, for $x \in B$. The effect of the collapse on the flowbox neighborhood is to take it to a collection of joining and splitting charts as per Definition 2.2.1 and Figure 2.4. The collapsing procedure may be done smoothly, yielding an ambient isotopy on finite links of periodic orbits.   □

This completes the proof of Theorem 2.2.4 in the case of a one-dimensional basic set. In this case, there are no "exceptional" orbits, as in the statement of Theorem 2.2.4 — the knots and (finite) links are in bijective isotopic correspondence.

Remark 2.2.8 Let us reformulate what we have done in terms of the symbolic dynamics. The flow restricted to the one-dimensional basic set B is conjugate to a suspended subshift of finite type. That is, any orbit can be put in 1:1 correspondence with a bi-infinite symbol sequence in $\Sigma_{\mathcal{A}}$, where $\mathcal{A}$ is the transition matrix for the subshift. In collapsing out the strong stable foliation, we are identifying orbits which asymptotically converge in forwards time. This has the effect of ignoring the past; hence, the template construction "chops off" the left half of every symbol sequence (the past), leaving a one-sided symbol sequence (the future). In particular, periodic orbits, whose pasts and futures coincide, are unaffected by this procedure. Orbits on a template can thus be put in bijective correspondence with a one-sided subshift of finite type (*cf.* Remark 1.2.27). We will return to this idea and consider it carefully in Section 2.4.

Exercise 2.2.9 Describe what happens, topologically and symbolically, when one collapses out an *unstable* foliation instead of a stable one. Does this always/necessarily yield the "same" template?

2.2.2   Case 2: the DA

Assume $\dim(\mathcal{B}) > 1$. We reduce this scenario to that of Case 1 by a procedure known as the *DA*, or, *derived from Anosov*. This modification to a flow is originally due to Smale [165], and has been explicitly described by Robinson [153], Franks and Robinson [57, Appendix], and Williams [190]. Synonymous terms for this construction include *Smale surgery* and *orbit splitting*. Our ultimate goal is, as in Case 1, to collapse M by identifying orbits in a strong stable foliation. But we cannot always do so directly:

Example 2.2.10 Let $f : T^2 \to T^2$ be the hyperbolic toral map of Example 1.2.7 and let ϕ_t be the suspension flow associated with f. This is a flow on the compact three-manifold $T^2 \times I/(x,0) \sim (f(x),1)$, which is not T^3 since f is not isotopic to the identity map. This flow has a hyperbolic chain-recurrent set; however, the dimension of the [unique] basic set is three (recall that typical orbits of f cover T^2 densely). If one nevertheless collapses each stable manifold to a point, the resulting space is *not* a template. Recall from Example 1.2.7 that stable manifolds of points under f wind about on T^2 densely. This implies that for the flow ϕ_t, the stable manifold of any point is arbitrarily close to that of any other point; hence, collapsing stable manifolds for this flow yields a non-Hausdorff space — certainly not the desired object.

The DA construction resolves this problem by first opening up a "hole" in M and separating the invariant manifolds.

Assume $\dim(\mathcal{B}) = 3$, and consider a closed orbit γ along with a small tubular neighborhood $N_\epsilon \equiv N_\epsilon(\gamma)$ of diameter ϵ. We will modify the flow ϕ_t on N_ϵ as follows. For each $x \in \gamma$, let $[e^s, e^u, e^c](x)$ be the coordinate frame based at the point x spanning the stable, unstable, and center directions (this is uniquely defined by the definition of hyperbolicity and by the Stable Manifold Theorem). For sufficiently small ϵ, the local planes spanned by e^s and e^u foliate N_ϵ with meridional discs. Consider the vector field X, given by:

$$X(x) = \begin{cases} (x_s, 0, 0) & x = (x_s, x_u, x_c) \in N_\epsilon \\ 0 & x \notin N_\epsilon \end{cases} . \tag{2.3}$$

The DA flow, ϕ_t^{DA}, is defined to be the flow generated by the vector field

$$\frac{d\phi^{DA}}{dt} = \frac{d\phi}{dt} + \kappa X, \tag{2.4}$$

for some $\kappa > 0$. The effect of adding κX is to "push out" the flow along the local stable manifold of γ. For very small κ, there is no qualitative change in the flow. But for κ larger than the contraction rate for the stable manifold of γ, the flow is altered .

Lemma 2.2.11 *For appropriate choice of increasing κ, γ bifurcates from a saddle-type orbit to a source along with one or two additional saddle-type orbits in a small tubular neighborhood of γ.*

Proof: Consider a local cross section Π for the flow, transverse to γ. Then, for $\kappa = 0$, γ is a fixed point under the induced return map. Consider further the cross section given by $I = W_{loc}^s(\gamma) \cap \Pi$ for I sufficiently long: this induces a hyperbolic return map r on the one-dimensional segment I. For $\kappa = 0$, the return map on I is a contraction by some factor $0 < \lambda < 1$ (*cf.* Theorem 1.2.9). Also, r may be orientation preserving or reversing, depending upon the orientation of the stable bundle E^s of γ.

Regard I as the interval $[-1, 1]$ with the fixed point corresponding to γ at the origin. Then, for $\kappa = 0$, the return map is conjugate to $x \to \pm \lambda x$, depending on whether the map is orientation preserving $(+)$ or reversing $(-)$. Increasing κ has the effect of changing the map on a small neighborhood of the origin, increasing the slope (in absolute value). At a certain $\kappa_* > 0$, there is a bifurcation when the slope at 0 is ± 1 (*cf.* §1.2.3). When r is orientation preserving, a pitchfork bifurcation occurs, since there is a symmetry $x \mapsto -x$ imposed. In this case, two new periodic orbits, γ' and γ'', are created, each isotopic to γ (though perhaps linked). In the nonorientable case, a period-doubling bifurcation occurs, creating a single orbit γ', isotopic to the *twisted double* of γ: see Figure 2.7. Each of the new orbits γ' and γ'' are of saddle-type, and γ has become a source (as per the description of §1.2.3). □

Versions of the following proposition appear in [165, 153, 190, 57].[1]

Proposition 2.2.12 *Let Λ denote the complement of $W^u(\gamma)$ for the DA flow ϕ_t^{DA} on $\mathcal{B}$. Then Λ is a hyperbolic expanding attractor.*

Proof: By definition,

$$W^u(\gamma) = \bigcup_{t>0} \phi_t^{DA}(W_{loc}^u(\gamma)); \tag{2.5}$$

---

[1] The results are proved only for the case of the toral Anosov diffeomorphism of Example 1.2.7.

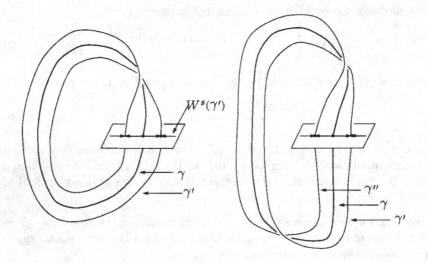

Figure 2.7: Orbit splitting creates one or two new saddle-type orbits.

hence, $\phi_t^{DA}(W_{loc}^u(\gamma)) \supset W_{loc}^u(\gamma)$ for $t > 0$ and the complement $\mathcal{B} \setminus W_{loc}^u(\gamma)$ is a positively invariant region for the flow. This implies that

$$\Lambda = \bigcap_{t>0} \phi_t^{DA}(M \setminus W_{loc}^u(\gamma)), \tag{2.6}$$

is an attractor. To show that Λ is hyperbolic, note first that from Equation (2.3), stable manifolds are preserved by the construction (except that of γ, of course): hence, the stable bundle E^s on Λ under ϕ_t^{DA} is precisely that of the original flow ϕ_t. Although the DA perturbation to ϕ_t disrupts the unstable bundle, E^u, it does so gently. To produce an unstable bundle on Λ, it suffices to construct *cones* in TM_x, for $x \in \Lambda$, whose sides are estimated from the effect of the DA perturbation on the unstable bundle of the original flow ϕ_t. Upon iteration, these cones converge to the new unstable bundle E^u. This is a procedure familiar to dynamicists: accounts and examples appear in [135, 76].

To show that Λ is expanding (recalling Definition 2.1.3), we first show that the complement, $W^u(\gamma)$, is dense in $\mathcal{B}$. Pick $\xi \in \mathcal{B}$. We claim that $W^s(\xi)$, the strong stable manifold of ξ under ϕ_t, is dense in B. Since B is a basic set, Theorem 1.2.23 states that there is a Markov partition for a cross-section of B with a continuous surjection from the subshift of finite type to the cross section of B. Hence, using the same trick as in Exercise 1.2.30, we can construct a symbolic stable manifold of ξ whose backwards orbit is dense in symbol-space. Then, since the map to B is a surjection, the stable manifold is dense.

However, the DA perturbation leaves the stable bundle invariant, so the stable manifold of ξ under ϕ_t^{DA} is also dense in $\mathcal{B}$. Choose $x \in \Lambda$ and N_x a small neighborhood in $\mathcal{B}$. Any $y \in N_x \cap W^s(\xi)$ flows by ϕ_t^{DA} arbitrarily close to any point in $\mathcal{B}$ in backwards time; However, this implies that $\phi_{-t}^{DA}(y)$ intersects

$W^u(\gamma)$ in the DA flow for t sufficiently large, since $W^u(\gamma)$ contains a tubular neighborhood of γ. Since $W^u(\gamma)$ is invariant under the flow, $y \in W^u(\gamma)$, which is thus arbitrarily close to $x \in \Lambda$.

As such, $W^u(\gamma)$ is dense in the three-dimensional basic set $\mathcal{B}$, so $\dim(\Lambda) \leq 2$. Consider the periodic orbit γ'. Since it is not in $W^u(\gamma)$, it must be a subset of Λ. Since Λ is an attractor, a small compact neighborhood N_Λ can be chosen which is forward invariant. Since $\gamma' \subset \Lambda$, it follows that $W^u_{loc}(\gamma') \subset N_\Lambda$. By definition, Λ is the intersection of the forward flow of N_Λ; thus, as the forward flow of $W^u_{loc}(\gamma')$ is the invariant manifold $W^u(\gamma')$, it follows that $W^u(\gamma') \subset \Lambda$. Since $W^u_{loc}(\gamma')$ is of topological dimension two, so is Λ.   □

Lemma 2.2.13 *With the exception of the additional orbits γ' and γ'', the periodic orbits of ϕ_t and those of ϕ_t^{DA} are in bijective isotopic correspondence.*

Proof: Let ϕ_t^ϵ denote the DA flow for a fixed tubular neighborhood N_ϵ of γ with diameter $\epsilon > 0$. Shrink ϵ continuously and consider the 1-parameter family of flows ϕ_t^ϵ as $\epsilon \to 0$. For each sufficiently small $\epsilon > 0$, the invariant set Λ_ϵ is hyperbolic. Hence, all the DA flows on Λ_ϵ for (small) $\epsilon > 0$ are topologically conjugate, and the 1-parameter family of homeomorphisms gives an isotopy between their periodic orbit sets. Since the DA flow is a modification of the original ϕ_t on the tubular neighborhood N_ϵ, those periodic orbits which do not intersect N_ϵ are identical, and hence isotopic. As $\epsilon \to 0$, every periodic orbit of ϕ_t eventually falls out of N_ϵ except γ, which is replaced in the DA by γ, γ', and (if necessary) γ''. □

Remark 2.2.14 By performing a DA splitting along γ, we have created one or two new orbits and reduced the topological dimension of our basic set to two. It is remarkable that a small perturbation to an Anosov flow can reduce the dimension of the basic set. One can picture this as follows: consider $W^u(\gamma)$ for the Anosov flow ϕ_t. This invariant manifold runs through M densely. After the DA perturbation, the creation of a source and two orbits γ' and γ'' may be thought of as "splitting" what was $W^u(\gamma)$ into a "thick" unstable manifold bounded by $W^u(\gamma')$ and $W^u(\gamma'')$. Thus, like thickening the rational points of an interval to obtain a Cantor set in the complement, the complement of $W^u(\gamma)$ in the DA flow is an attractor which is locally the product of $D^2 \times C$, where C is a Cantor set.

Remark 2.2.15 From the work of Williams on expanding attractors [192], it follows that the attractor Λ is transitive: a basic set.

We may attain our goal of reducing the dimension of the basic set to one by performing another splitting on another closed orbit. Suppose Λ is a basic set of dimension two. Since Λ is two dimensional and hyperbolic and M three-dimensional, the stable, unstable, and center bundles must each be of dimension one. Since Λ must contain the center bundle, it must also contain either the stable or unstable bundle, leaving only the remaining direction. Hence, Λ is either an attractor or a repellor.

Assume Λ is a repellor (this is the opposite of what one obtains from a DA on a three-dimensional basic set, but one may reverse time and so obtain a repellor). Then, as before, choose a closed orbit $\hat{\gamma}$ (if applicable, one of the "new" orbits obtained from the DA would do nicely) and modify the flow on a small neighborhood as in Equations (2.3) and (2.4). As before, this creates one or two new saddle-type orbits in the new basic set, $\hat{\gamma}'$ and $\hat{\gamma}''$, while changing $\hat{\gamma}$ to a source.

Let $\hat{\Lambda}$ denote the complement of $W^u(\hat{\gamma})$ in Λ. The arguments of Proposition 2.2.12 carry over almost verbatim to show that $\hat{\Lambda}$ is a basic set of dimension one. The steps proceed as follows, with details as in Proposition 2.2.12:

1. $\hat{\Lambda}$ is hyperbolic: orbit splitting leaves stable bundles invariant — estimate unstable bundles via cones.

2. $W^u(\hat{\gamma})$ is dense in $\hat{\Lambda}$: arguing as in Proposition 2.2.12.

3. $\dim \hat{\Lambda} = 1$: since $W^u(\hat{\gamma})$ is dense in the two-dimensional Λ, $\dim \hat{\Lambda} \leq 1$, but $\hat{\Lambda}$ contains one-dimensional flowlines.

Also, as in Lemma 2.2.13, the periodic orbit set is unchanged except for the additional orbits $\hat{\gamma}'$ and $\hat{\gamma}''$ since we modify the flow on an arbitrarily small neighborhood of an orbit.

Proof of Theorem 2.2.4: After at most two orbit splittings, one may reduce the basic set B to the one-dimensional Case (1); then, by collapsing out a strong stable foliation, the desired template is obtained.   □

Remark 2.2.16 In the case of the orbit splitting involved in the DA construction, one begins with a knot γ and replaces it with either two isotopic copies of itself (perhaps linked), or with a "doubled" knot (perhaps twisted). Since there are at most two orbit splittings, there are at most two extraneous knots in the template which do not correspond to closed orbits in the original flow. Note, however, that any closed orbit is suitable for splitting; different choices may yield ostensibly different templates.

Remark 2.2.17 A version of Theorem 2.2.4 in higher dimensions would be desirable. There are impassable obstructions to this, not the least of which is the fact that knotting and linking of orbits in dimensions higher than three is nonexistent. In addition, the orbit-splitting procedure is more dramatic in higher dimensions, where, instead of creating one or two additional orbits (an S^1 bundle over S^0), an entire S^1 bundle over S^k is created in dimension $k + 3$. Of course, under unusual circumstances, a high-dimensional flow contains global strongly contracting directions which allow one to first reduce to a three-dimensional flow and then proceed as usual; however, the original flow is not then essentially high dimensional.

Remark 2.2.18 Several authors have used branched two-manifolds of a slightly different form than the templates of this chapter – these are *closed* (boundaryless) branched two-manifolds. The definition in terms of charts is slightly different (see [192, 37]), but a closed branched two-manifold can usually be transformed into a template via splitting along a finite number of orbits. These branched manifolds have been used to characterize hyperbolic attractors in flows [192, 37] as well as to capture incompressible surfaces in three-manifolds [82, 48, 61].

2.3   Examples and applications

In this section, we present a collection of examples of templates, along with typical situations in which one may use templates to capture the periodic orbits in a flow or a portion of a flow. The following subsections include a variety of topics, from ODEs to fibred 3-manifolds to time series. Though we will refer back to several of these examples in subsequent chapters, the reader may skip or skim the following without serious loss of continuity.

2.3.1   The Lorenz-like templates

Example 2.3.1 (Lorenz-like templates) The simplest examples of templates are those formed from a single branch line chart with two strips: the *Lorenz-like templates*. For $m, n \in \mathbb{Z}$, denote by $\mathcal{L}(m, n)$ the template pictured in Figure 2.8(a). The two unknotted, unlinked strips have m and n signed half-twists respectively.

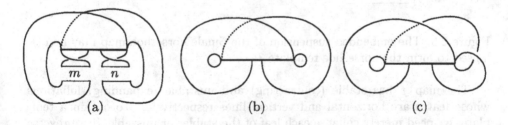

Figure 2.8: (a) The Lorenz-like template $\mathcal{L}(m, n)$; (b) the Lorenz template $\mathcal{L}(0, 0)$; (c) the horseshoe template $\mathcal{L}(0, 1) = \mathcal{H}$.

Example 2.3.2 The Lorenz template, $\mathcal{L}(0, 0)$, is pictured in Figure 2.8(b). This template is an idealization of the attractor for Equation (2.1) in Example 2.0.1. The link of periodic orbits supported on $\mathcal{L}(0, 0)$ has a number of interesting properties, as shown by Birman and Williams [23]. We list some of these properties here and refer the reader to [23] and [195] for proofs.

Theorem 2.3.3 (Birman and Williams [23], Williams, [195]) *Let L be a link of $\mu \geq 1$ components on $\mathcal{L}(0,0)$. Then L is a positive braid and also a fibred link (see Definition 2.3.10). Every component of L is prime. Every torus knot lives on $\mathcal{L}(0,0)$.*

Example 2.3.4 (the horseshoe template) The horseshoe template, $\mathcal{H}$, is isotopic to the Lorenz-like template $\mathcal{L}(0,1)$ of Example 2.3.1. However, the method of obtaining this template from Smale's horseshoe map (Example 1.2.28) is crucial.

Recall from Example 1.2.28 that the standard horseshoe map f acts on a square $I^2 \subset \mathbb{R}^2$, depicted in Figure 2.9. Suspending f yields a flow on a mapping torus $I^2 \times S^1$. Embedding this flow into $\mathbb{R}^3$ in the "standard" way (no additional twists) yields a well-defined suspension flow as depicted in Figure 2.9. Since f is hyperbolic, the conditions of Theorem 2.2.4 are satisfied and we may obtain a template, $\mathcal{H}$.

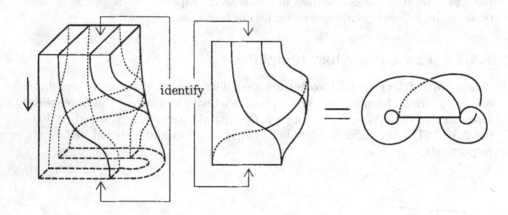

Figure 2.9: The embedded suspension of the Smale horseshoe map may be collapsed to form the horseshoe template $\mathcal{H}$.

The map f has stable (contracting) and unstable (expanding) foliations whose leaves are horizontal and vertical lines respectively. To obtain a template, we need merely collapse each leaf of the stable (or unstable, if we reverse time) foliation to a point. This appears in Figure 2.9 also, where the resulting template $\mathcal{H}$ is seen to be isotopic to the Lorenz-like template $\mathcal{L}(0,1)$.

Holmes and Williams [93] and Holmes [88, 90] have made extensive studies of which types of knots live on the template $\mathcal{H}$: see [70] for a review. We will use the horseshoe template in Chapter 4 to derive more general results for bifurcations in ODEs. In contrast to Theorem 2.3.3, the following proposition will be proved in §4.2 concerning knots on $\mathcal{H}$:

Proposition 2.3.5 (Holmes and Williams [93]) *The horseshoe template $\mathcal{H}$ contains no (p,q) torus knots for which $p < 3q/2$ (or, equivalently, $q < 3p/2$).*

In general, little is known about which knots live on the Lorenz-like templates for arbitrary m, n — even for such a simple family as torus knots. But perhaps knowing something about which knots live on some $\mathcal{L}(m, n)$ gives information about the existence of this knot on other Lorenz-like templates.

Problem 2.3.6 *For which pairs of integers* (m, n) *and* (m', n') *is it true that any knot which lives on* $\mathcal{L}(m, n)$ *must also live on* $\mathcal{L}(m', n')$?

Sullivan [168] has given a partial answer to this question. We will return to Problem 2.3.6 and fill in some of the gaps later in §3.2 and §3.3.

2.3.2 Nonlinear oscillators, horseshoes, and Hénon maps

In this and the following subsection, we indicate how hyperbolic sets and templates such as those introduced above arise in some specific classes of flows and maps.

Versions of the Smale horseshoe (Example 1.2.28) can appear naturally in periodically forced oscillators of the form

$$\ddot{x} = f(x, \dot{x}, t) \quad ; \quad f(x, \dot{x}, t) = f(x, \dot{x}, t + T), \tag{2.7}$$

for fixed $T > 0$. Letting $\dot{x} = y$, $t = \theta$, and regarding θ as an element of $S^1 = \mathbb{R}^1 / T\mathbb{Z}$, we may rewrite (2.7) as a vector field on a two-manifold cross S^1:

$$
\begin{aligned}
\dot{x} &= y \\
\dot{y} &= f(x, y, \theta) \\
\dot{\theta} &= 1.
\end{aligned}
\tag{2.8}
$$

Example 2.3.7 We give two examples of forced oscillators as per Equation (2.8): the Duffing equation,

$$
\begin{aligned}
\dot{x} &= y \\
\dot{y} &= x - x^3 - \delta y + \gamma \cos \omega \theta \qquad (x, y, \theta) \in \mathbb{R}^1 \times \mathbb{R}^1 \times S^1 \\
\dot{\theta} &= 1;
\end{aligned}
\tag{2.9}
$$

and the forced, damped pendulum,

$$
\begin{aligned}
\dot{\phi} &= v \\
\dot{v} &= -\sin \phi - \delta v + \gamma_0 + \gamma_1 \cos \omega \theta \qquad (\phi, v, \theta) \in S^1 \times \mathbb{R}^1 \times S^1 \\
\dot{\theta} &= 1.
\end{aligned}
\tag{2.10}
$$

Here, δ, γ, ω, etc. are parameters which may be varied externally to induce bifurcations in the flows. These and other examples arise in physics and engineering as models of mechanical and electrical devices (*e.g.*, [137, 4]). In the case of Equation (2.9), uniformly bounded solutions such as periodic orbits live within a compact region $D^2 \times S^1$ of the phase space; in the case of Equation (2.10), the appropriate region is $S^1 \times I^1 \times S^1 = A \times S^1$, where A denotes the annulus.

In general, a global cross section $\Pi = \{(x, y, \theta) : \theta = 0\}$ exists on which the flow of (2.8) induces a Poincaré map, P. For both equations (2.9) and (2.10), with positive damping $\delta > 0$,

$$\det DP = \exp\left(\int_0^T \text{trace}\,[\text{Jacobian}(P)]\,dt\right) = e^{-\delta T}, \qquad (2.11)$$

so P uniformly contracts areas, and there is a compact trapping region (D^2 or A, in these cases) into which *all* orbits eventually enter and thereafter remain, and which contains the attractor. See, for example, [76, 85]. For specific ODEs, such as those above, for small damping (δ) and forcing (γ), certain perturbation methods, pioneered by Melnikov [120], may be used to prove the existence of transverse homoclinic orbits to a hyperbolic periodic orbit: see Figure 2.10(a) and [76]. Then, by Theorem 1.2.33, there exists a Smale horseshoe within the return map. More precisely, some iterate P^N of P contains a full shift on two symbols. In the simplest case, $N = 1$, and, as indicated in Figure 2.10(b), for the Duffing equation, we have precisely the suspension of the horseshoe given in Figure 2.9. More complicated embeddings of the horseshoe template within a forced oscillator are, of course, abundant in cases where $N > 1$.

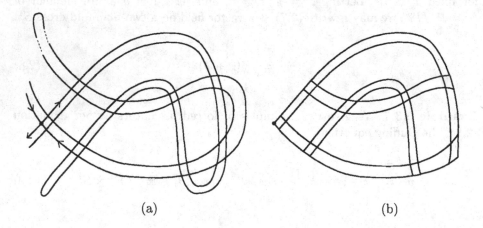

(a)   (b)

Figure 2.10: A Poincaré map for the forced Duffing equation; (a) invariant manifolds; (b) the "simplest" horseshoe.

While properties of such Poincaré maps, including the existence of homoclinic orbits, can be proven, explicit expressions for these maps cannot be obtained. Consequently, much in the spirit of Guckenheimer's and Williams's construction of a geometrical Lorenz attractor [77], Hénon, in 1976 [83], proposed a polynomial mapping that models the behavior of the Smale horseshoe.[2] This

---

[2]He actually did this in connection with the Lorenz equation in a different parameter regime from (2.1).

two-parameter family may be written

$$(x, y) \mapsto (y, -\epsilon x + \mu - y^2). \tag{2.12}$$

(A different, albeit equivalent form appears in [83].) Observe that $\det DF = \epsilon$ is constant, so that, for $0 < \epsilon < 1$ the map preserves orientation and contracts area uniformly, as do the Poincaré maps discussed above. For $\epsilon = 1$, it preserves area, and for $\epsilon = 0$, all orbits collapse in one iterate to the parabola $y = \mu - x^2$, after which their behavior is governed by the one-dimensional map

$$y \mapsto \mu - y^2, \tag{2.13}$$

mentioned in §1.2.3.

For large μ $\left[\mu > \left(\frac{5 + 2\sqrt{5}}{4}\right)(1 + |\epsilon|^2) \text{ suffices } [42]\right]$, (2.12) contains a full shift on two symbols, while for $\mu < \frac{1}{4}(1 + \epsilon)^2$, the chain-recurrent set is empty. For fixed ϵ and increasing μ, an infinite sequence of bifurcations occurs in which the horseshoe, with its countable set of periodic orbits, is created. The Hénon map provides a useful model for horseshoe creation, to which we shall return in §4.2. In fact, it has recently been shown that the Hénon map with small ϵ is present in an asymptotic limit for high iterates of *all* Poincaré maps near the (global) bifurcations in which homoclinic orbits are created in quadratic tangencies [140, 131].

Due to the first component of the vector field (2.8), the maps considered above preserve orientation and derive from, or lead naturally to, flows with orbit crossings all of one sign, hence yielding positive templates. In the next subsection, we introduce a class of flows which yield more general templates.

2.3.3 Shil'nikov connections

Recall the Poincaré-Birkhoff-Smale Theorem (Theorem 1.2.33), which we used in Section 1.2.2 to embed horseshoe-like templates within a three-dimensional flow containing a transverse homoclinic orbit to a periodic orbit. The next family of examples we consider is derived from a similar theorem, due to L. P. Shil'nikov, which proves the existence of suspended horseshoes near certain types of homoclinic connections to a fixed point:

Definition 2.3.8 A *Shil'nikov connection* for a flow ϕ_t on $\mathbb{R}^n$ ($n \geq 3$) is an orbit Γ which satisfies the following two conditions:

1. Γ is *homoclinic* to a hyperbolic fixed point p, and Γ must be bounded away from all other fixed points.

2. The linearization $D\phi|_p$ of the flow at p has leading eigenvalues $\{-\lambda^s \pm \omega i, \lambda^u\}$, with

$$\lambda^u > \lambda^s > 0 \qquad \omega \neq 0. \tag{2.14}$$

By "leading" is meant that any other eigenvalues have real parts outside of the interval $[-\lambda^s, \lambda^u]$.

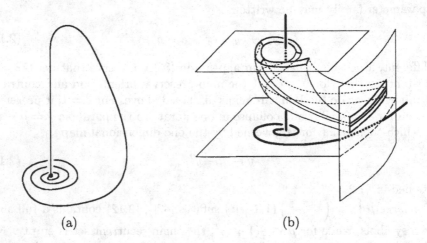

$$(a) \hspace{5cm} (b)$$

Figure 2.11: (a) A Shil'nikov connection in $\mathbb{R}^3$; (b) the Markov partition for a suspended horseshoe.

Shil'nikov connections occur frequently in systems modeling physical phenomena, such as flow through pipes [36], coupled oscillators [187], magnetoconvection [155], and electric circuits [38, 105]. The following theorem was first proved by Shil'nikov [160, 161], with extensions and repetitions later in [179] and elsewhere. A number of textbooks also contain these results along with proofs [76, 188, 189].

Theorem 2.3.9 (L. P. Shil'nikov [161]) *Let ϕ_t be a flow supporting a Shil'nikov connection Γ to a fixed point p. Then, there exists a countable infinity of suspended Smale horseshoes in the flow in an arbitrarily small tubular neighborhood of the homoclinic orbit Γ. Under a small C^1 perturbation, finitely many of these horseshoes remain.*

We give an outline of the proof of Theorem 2.3.9 in §4.4.2.

The entire flow near Γ does not satisfy the hyperbolicity requirements of Theorem 2.2.4: moreover, there are numerous features of the dynamics and (especially) bifurcations of flows near such orbits that are still poorly understood. However, the individual horseshoes implied by Theorem 2.3.9 *are* hyperbolic, and if, as in the previous subsection, we restrict our attention to any such subset of the flow, we may employ Theorem 2.2.4 to obtain a template which captures a *portion* of the flow, concluding that orbits on the embedded horseshoe templates are in one-to-one isotopic correspondence with a proper subset of orbits in the flow near Γ. This is our strategy for finding templates within this class of flows. The task, then, is to carefully track how the suspended horseshoes are embedded within the flow.

The proof of Theorem 2.3.9 involves constructing Poincaré sections transverse to Γ near the fixed point p and linearizing the flow near p and along Γ to obtain approximate return maps. The horseshoes are constructed by flowing pairs of rectangles near p and then along Γ: see Figure 2.11.

Since these horseshoes are hyperbolic, we can keep track of their stable and unstable foliations. By collapsing one set of these foliations and carefully following the embedding, we construct an embedded template. First, we collapse the flow near the fixed point p, yielding two strips which, due to the spiraling nature of the flow, wind about $W^u(p)$ in N full twists before fusing at a branch line: see Figure 2.12(a). Secondly, we follow the template along the unstable manifold $W^u(p)$, twisting an unspecified number of times along with the stable/unstable bundles of $W^u(p)$ before reconnecting: see Figure 2.12(b). (The number depends upon the size of the neighborhood of p on which the local, almost-linear, map is constructed: the neighborhood must be taken sufficiently small for various cone estimates, necessary for hyperbolicity, to hold.) Assuming that $W^u(p)$ is unknotted, this construction yields an embedding of the template obtained by inserting a finite number of half-twists in the horseshoe template $\mathcal{L}(0,1)$ after the branch line.

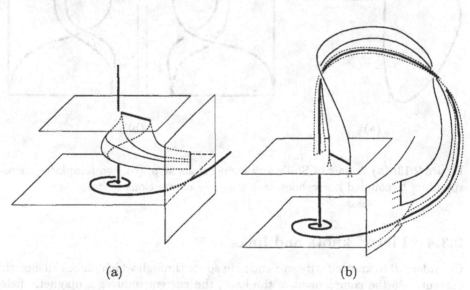

(a)   (b)

Figure 2.12: (a) The template near the fixed point p; (b) global twisting along the unstable manifold.

The fact that there are an indeterminate number of twists in the above template is a difficulty: given a system containing a Shil'nikov connection, it is known only that these templates exist in the flow for sufficiently large amounts of twisting. We will address this later in §4.4, after developing more tools.

Despite the apparent indeterminacy of these templates, they exhibit several interesting features. For example, all of the suspended horseshoes near the homoclinic orbit are disjoint and link one another in various ways. In addition, a number of extensions to Theorem 2.3.9 exist [179]: besides suspended horseshoes, there are also suspended full N-shifts for any $N > 0$. Hence, a variety of complicated templates are embedded in these flows, which capture (portions of) the periodic orbit set. Finally, when the vector field is symmetric or when two-parameter families are considered, there is the possibility of a fixed point p supporting a *pair* of Shil'nikov connections. Such a structure might appear as in Figure 2.13(a). The appendix of [71] catalogues the possible templates in these situations.

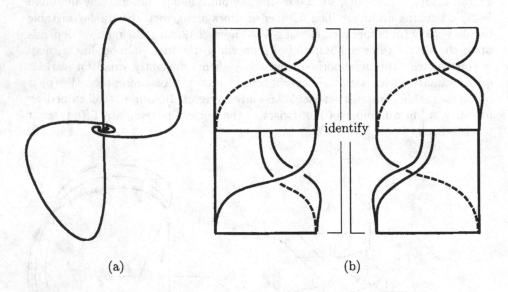

(a)   (b)

Figure 2.13: (a) A pair of Shil'nikov connections at p; (b) two templates corresponding to coupled horseshoes near a pair of connections.

2.3.4   Fibred knots and links

Consider a thin knotted wire suspended in space through which passes an electric current. On the complement of the knot, the current induces a magnetic field which may have closed field lines. The way in which these closed curves entwine the wire is intimately related to the knotting of the wire. This concept of an induced field on the compliment of a knot is made mathematically precise by the notion of a fibred knot.

A knot or link K in S^3 is *fibred* if the complement $S^3 \setminus K$ fibres over S^1 with fibre a Seifert spanning surface M [154, 33]. More specifically,

Definition 2.3.10 A knot or link K is *fibred* if there exists an orientable surface M with boundary $\partial M = K$ and a homeomorphism $\Phi : M \to M$ such that the complement $S^3 \setminus K$ is homeomorphic to the quotient space $(M \times [0,1])/ \sim$ where $(x,0) \sim (\Phi(x),1)$. The surface M is the *Seifert spanning surface* and the map Φ is the *monodromy*.

The simplest example of a fibred knot is the unknot, which has as fibre the disc D^2 and monodromy the identity map $\mathrm{id} : D^2 \to D^2$. Figure 2.14 illustrates the fibration of the complement in S^3, where it is seen that a fibration is akin to "blowing a bubble" M with bubble-ring K so as to fill out all of the complement, through the point at infinity, returning to the initial configuration. In Figure 2.14, each disc has the unknot as its boundary — we have cut open some of the discs for visualization.

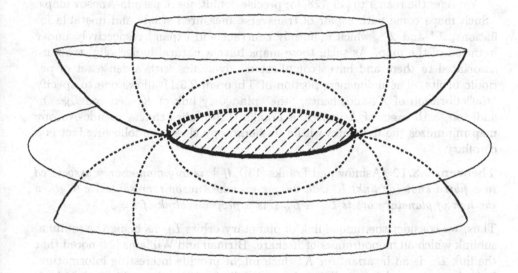

Figure 2.14: The fibration of the unknot complement by discs.

In fibring the complement in this manner, a flow is induced on $S^3 \setminus K$ by following a point on M as it is pushed through the complement. This is precisely the suspension flow of the monodromy Φ embedded in $S^3 \setminus K$. The monodromy Φ is thus a global return map for the flow, defined on the Seifert surface M, which completely captures the dynamics. Alternatively, there exists a map $\pi : S^3 \setminus K \to S^1$, called the *fibration*, which has as its fibre $\pi^{-1}(\theta)$ for $\theta \in S^1$ an embedded copy of M. Then the flow on the complement is precisely the integration along the gradient of the fibration $\pi : S^3 \setminus K \to S^1$.

Any periodic points of the monodromy Φ become periodic orbits of the suspension flow which coil about the base knot K in a manner determined by the fibration. The resulting collection of knots was dubbed, by Birman and Williams [24], the *planetary link* for K with monodromy Φ: $L_{K,\Phi}$.

Since M is a surface and Φ a diffeomorphism, one may invoke the Nielsen-Thurston classification of surface diffeomorphisms [178, 46]:

Theorem 2.3.11 (Nielsen [138], Thurston [177]) *A surface diffeomorphism $\Phi : M \to M$ is isotopic to a unique homeomorphism $\hat{\Phi}$ such that one of the following holds:*

1. $\hat{\Phi}$ is periodic, i.e., $\hat{\Phi}^k = \mathrm{id}$ for some k;

2. $\hat{\Phi}$ is pseudo-Anosov (see below); or

3. $\hat{\Phi}$ is reducible, i.e., there exists an invariant family $\mathcal{C}$ of disjoint simple closed curves on M such that the restriction of Φ to the complement of $\mathcal{C}$ decomposes into a finite number of disjoint maps which are either periodic or pseudo-Anosov.

We refer the reader to [46, 178] for precise definitions of pseudo-Anosov maps. Such maps come with a pair of transverse measured stable and unstable foliations, $\mathcal{F}^s$ and $\mathcal{F}^u$, which uniformly contract and expand respectively under iteration of the map. As such, these maps have a natural hyperbolic structure associated to them and have "complicated" dynamics with a dense set of periodic orbits. The uniqueness portion of Theorem 2.3.11 allows one to specify "the" fibration of K, and, hence, "the" planetary link of K, denoted L_K. In addition, a theorem of Asimov and Franks [13] implies that a pseudo-Anosov map minimizes the dynamics within its homotopy class: the following fact is a corollary.

Theorem 2.3.12 (Asimov and Franks [13]) If Φ is any monodromy associated to a fibred knot (or link) K with unique pseudo-Anosov representative $\hat{\Phi}$, then the link of planetary orbits $L_K \equiv L_{K,\hat{\Phi}}$ is a proper sublink of $L_{K,\Phi}$.

Thus, we consider the unique link of planetary orbits L_K as being the minimal sublink which all monodromies of K share. Birman and Williams [24] noted that the link L_K is an invariant for K which might provide interesting information. In their study of planetary links, they carefully considered the figure-eight knot (see Figure 1.1(c)), which is fibred with fibre a punctured torus and monodromy isotopic to the Anosov map of Example 1.2.7,

$$\hat{\Phi} = \begin{bmatrix} 2 & 1 \\ 1 & 1 \end{bmatrix}, \tag{2.15}$$

acting on the universal cover $\mathbb{R}^2 \setminus \mathbb{Z}^2$ [33, p. 73].

Because the pseudo-Anosov map $\hat{\Phi}$ satisfies the hyperbolicity requirements of Theorem 2.2.4, it is possible to collapse the complement of the figure-eight knot down to a template. Birman and Williams, in [24], derive two templates for the fibration of the complement of the figure-eight knot (corresponding to $\hat{\Phi}$) — one via direct visualization, and the other indirectly by means of branched coverings of S^3: we recall their templates in Figure 2.15.

Of course, since the map $\hat{\Phi}$ of Equation (2.15) is Anosov, the DA process of §2.2.2 must be performed; hence, there may be two extraneous orbits on the template not present in the original flow.

Simple fibred knots and links in S^3 often (if not always) give rise to very complicated templates supporting their planetary links. The *Whitehead link*, L_W, displayed in Figure 2.16, is a fibred link with pseudo-Anosov monodromy. Using the techniques in [24], we have shown that the planetary link for L_W is supported on the template illustrated in Figure 2.17.

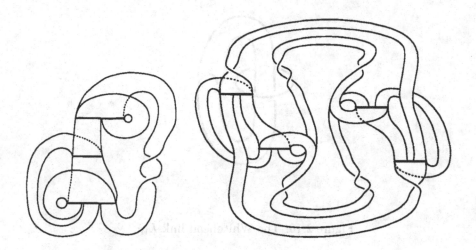

Figure 2.15: The "direct" and "indirect" versions of the figure-eight template.

2.3.5 Templates from time series

Finally, we consider a class of examples about which little is known rigorously, but which may have important applications, particularly for experimentalists seeking geometrical models of dynamical processes. Consider an experimental measurement of a continuous scalar variable whose dynamical behavior is complicated: *e.g.*, a temperature reading, a chemical concentration, or a speed. The data is received in the form of a *time series*: a function $\rho : [0, T] \to \mathbb{R}$, where T is the length of the data segment (in units of time).

Given a complicated time series, one would wish (among other things) to extract the essentials of the underlying dynamics. For example, consider a typical orbit of the Lorenz system (Equation (2.1)), and let $\rho(t)$ denote the projection of this orbit onto one of the coordinates (see Figure 2.18). Over long periods, this might appear to be without coherent form; yet, given its origins, there is certainly structure within the data. One is more suspicious of, say, the Dow Jones average, hiding some covert pattern.

Typically, one employs a variety of means for accessing "hidden" dynamical information within a time series: Fourier spectral content, statistical measures, fractal dimensions, and other tools provide certain types of information, while ignoring other, more geometric data. Fortunately, a theorem of Takens [175] suggests that one can often embed an attractor into a low-dimensional manifold via a "time delay" function, capturing the geometric and topological properties:

Theorem 2.3.13 (Takens [175]) *Let M be a compact n-manifold with a C^2-flow ϕ_t and a C^2-function $\tau : M \to \mathbb{R}^1$. Then, generically, the time-delay mapping $\Phi : M \to \mathbb{R}^{2n+1}$ defined by*

$$\Phi(x) = (\tau(x), \tau(\phi_1(x)), \tau(\phi_2(x)), \ldots, \tau(\phi_{2n}(x))) \qquad (2.16)$$

Figure 2.16: The Whitehead link L_W.

is an embedding.

A topological perspective has been proposed by Mindlin, Solari, Gilmore, Tufillaro, *et al.* [128] (*cf.* [180]), in which knot and link types of periodic orbits in the embedded flow are computed and related to a template. We outline the procedure detailed in [128].

1. Given a "chaotic" time series $\rho(t)$, extract a finite collection of low-period unstable periodic orbits, $\{\gamma_i\}_1^N$. This is done by examining "close returns" within the data, which are assumed to wander back and forth among many unstable periodic orbits. The low-period orbits are easiest to spot.

2. Map the time series into $\mathbb{R}^3$ via the (Takens) time-delay function, and assume that it is an embedding. There are several ways to realize this via different "filters" of the data. Clearly, this may not be possible in general: for success, orbits must appear to lie on a topologically two-dimensional attractor.

3. Consider the (small) collection $\{\gamma_i\}$ of low-period unstable periodic orbits computed in step (1). Embed these in $\mathbb{R}^3$ as per the embedding of step (2). Calculate their knot types, linking numbers, and self-linking numbers (*i.e.*, twisting of the stable/unstable bundles). These form a *basis* for the *induced template*.

4. Let $\mathcal{T}_\rho$ denote the "simplest" template in $\mathbb{R}^3$ which contains the basis $\{\gamma_i\}$. For example, if a global cross section to the flow exists, $\mathcal{T}_\rho$ is a template consisting of one branch line such that each γ_i lives on $\mathcal{T}_\rho$ and crosses the branch line the same number of times as the period of γ_i in the return map of the flow. The knot types, linking numbers, and self-linking numbers tell one how the strips of $\mathcal{T}_\rho$, each of which contains at least one γ_i, are knotted, linked, and twisted, respectively.

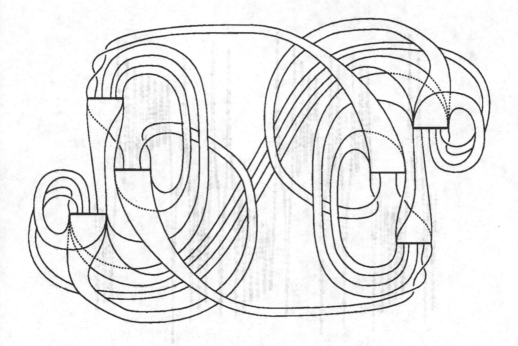

Figure 2.17: The Whitehead template.

After producing the induced template $\mathcal{T}_\rho$ for the data set ρ, one may now proceed to verify that the template $\mathcal{T}_\rho$ provides an accurate model of the dynamics. This can be done in a number of ways: *e.g.*, find higher-period orbits in the data set and confirm that these live in $\mathcal{T}_\rho$ with the appropriate embedding, or take another data set, ρ', and compute an induced template for this set.

When the induced template construction is successful, there are a number of benefits both to the experimentalist and to the theorist hoping to model the experiment from which it derives. First, an induced template offers a certain degree of prediction — one may identify a periodic orbit in the template, then go "hunting" for it in the data set. A successful example of this is documented in [128]. Secondly, one may verify models of the system. Should one model the experimental system with a set of ODEs, one takes a time series of the ODE solution and constructs the induced template for this data set. If the induced template for the model differs from the induced template for the experiment, this may indicate a shortcoming in the model.

There are, however, serious questions concerning this approach. Experimental systems are rarely three-dimensional and hyperbolic; hence, the use of templates to model them is, at the very least, suspect. In addition, the only guiding principal behind the choice of the induced template is Occam's Razor. As such, it is not surprising that many of the induced templates computed in practice are isotopic to the horseshoe template, $\mathcal{L}(0,1)$, or its mirror image [128, 180]

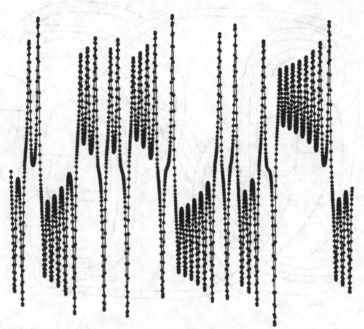

Figure 2.18: A time series derived from the Lorenz equations.

(though see [106] for an exception).

These doubts notwithstanding, there are numerous open questions about the use of induced templates for time series, whose answers could be of great value to experimentalists and modelers of complicated dynamics.

2.4   A symbolic language

Much of this book is concerned with templates and the links they carry. To analyze these, it is often useful to extract *subtemplates*, or subsets which are themselves templates (see Definition 2.4.6). In the late eighties, one of us [MS] noticed that the template $\mathcal{V}$, illustrated in Figure 2.21 below, contains a subtemplate which is isotopic to itself: see Figure 2.22 (this was used to show the existence of highly-composite knots on $\mathcal{V}$ [169]). In this section, we introduce conventions for symbolic descriptions of orbits and templates, which enables us to significantly generalize this kind of procedure to cases in which direct visualization is not possible.

2.4.1   Markov structures and symbolic coordinates

Recall from the proof of the Template Theorem in §2.2 that there is a natural correspondence between orbits which remain on a template and one-sided symbol sequences in a subshift of finite type: in particular, following upon Remark 2.2.8, we have

Lemma 2.4.1 *Given a template $\mathcal{T}$, label the strips $\{x_i : i = 1..N\}$. Let $\mathcal{A}_{\mathcal{T}}$ be an $N \times N$ matrix with entry $\mathcal{A}_{\mathcal{T}}(i,j) = 1$ if the incoming portion of x_i meets the outgoing portion of x_j at a branch line, zero otherwise. Then $\Sigma_{\mathcal{T}}$, the set of all forward orbits which remain on $\mathcal{T}$, is precisely the set of admissible sequences in the subshift of finite type given by $\mathcal{A}_{\mathcal{T}}$.*

Proof: See the proof of Theorem 2.2.4, or simply collapse $\mathcal{T}$ along the transverse direction of the semiflow, reducing $\mathcal{T}$ to an oriented graph. Then the orbits on $\mathcal{T}$ are one-sided directed paths on this graph: *cf.* Remark 1.2.22.   □

The way in which orbits fit together on a template $\mathcal{T}$ is described by placing a coordinate system on the branch lines $\{\ell_j : j = 1..M\}$, following the kneading theory of §1.2.3, and specifying the induced coordinates on $\Sigma_{\mathcal{T}}$. This ordering of orbits on a template is a key ingredient in discerning the relative placement of orbits on a template which might be too complicated to visualize.

Definition 2.4.2 Let $\mathcal{T}$ be a template with strips labeled $\{x_i\}_1^N$. Denote by $\{\ell_j\}_1^M$ the branch lines of $\mathcal{T}$ (one for each branch line chart). Then $\Sigma_{\mathcal{T}}$ is partitioned into N *branch segments*, denoted $\{\beta_i(\mathcal{T})\}_1^N$, where

$$\beta_i(\mathcal{T}) \equiv \{\mathbf{a} = a_0 a_1 a_2 \ldots \in \Sigma_{\mathcal{T}} : a_0 = x_i\}. \tag{2.17}$$

Denote by $\Sigma_{\ell_j} \subset \Sigma_{\mathcal{T}}$ the union of $\beta_i(\mathcal{T})$ over all i such that the strip x_i emanates from the branch line ℓ_j. We will sometimes refer to the union of the $\beta_i(\mathcal{T})$ as the *branch set*, denoted $\beta(\mathcal{T})$.

Proposition 2.4.3 *There exists a total ordering $\lhd$ on each Σ_{ℓ_j} which respects the topology of Σ_{ℓ_j}: that is, if $\mathbf{a} \lhd \mathbf{b}$ and $\{\mathbf{a}_n\}$ is a sequence converging to $\mathbf{a}$ then, for sufficiently large n, $\mathbf{a}_n \lhd \mathbf{b}$.*

Proof: This follows from the kneading theory [125], as outlined in §1.2.3. We construct $\lhd$ explicitly in what follows, and it will be seen to have the following property: $\lhd$ is the total ordering induced by the one-dimensionality of ℓ_j. That is, any point of an ℓ_j is an orbit which "begins" on ℓ_j. Orienting ℓ_j yields a total order on Σ_{ℓ_j} which respects the topology.   □

For the moment, assume $\mathcal{T}$ is an orientable template. Each branch line ℓ_j is one-dimensional. Hence, the set of branch segments in each ℓ_j are ordered (up to orientation of ℓ_j). If, for example, the branch segments $x_1, x_2, \ldots, x_p$ lie in ℓ_1 in this order, then choose $\lhd$ as either

$$x_1 \lhd x_2 \lhd \ldots \lhd x_p, \quad \text{or} \quad x_p \lhd \ldots \lhd x_2 \lhd x_1. \tag{2.18}$$

Having chosen an orientation for each ℓ_j, one then orders each Σ_{ℓ_j} lexicographically with respect to the ordering on the generators $\{x_i\}$. That is, given $\mathbf{a}$ and $\mathbf{b} \in \Sigma_{\ell_j}$, let J equal the index of the first symbol in which $\mathbf{a}$ and $\mathbf{b}$ disagree:

$$J = \min\{j : a_j \neq b_j\}. \tag{2.19}$$

Then $\mathbf{a} \lhd \mathbf{b}$ if $a_J \lhd b_J$, else $\mathbf{b} \lhd \mathbf{a}$. Of course, one cannot compare points in different Σ_{ℓ_j}: there is no notion of orientation for points on disjoint branch lines. Since $\mathcal{T}$ is orientable, the lexicographical ordering of itineraries corresponds to the ordering on the branch lines and it yields a natural coordinate system.

For nonorientable templates, the issue is no more difficult, but it does demand more bookkeeping. If a particular strip, say x_j, contains an odd number of half-twists (*i.e.*, the return map is orientation reversing on that interval), then one must keep track of the parity of that symbol in using $\lhd$ as in the invariant coordinate construction for the one-dimensional map f_T of §1.2.3.

Specifically, given a nonorientable template $\mathcal{T}$, construct a provisional ordering $\tilde{\lhd}$ as for an orientable template induced by the ordering on the individual branch lines (as above). This ordering $\tilde{\lhd}$ does not, however give an ordering on $\mathcal{T}$ which respects the topology of the branch lines. Now, given some planar presentation of $\mathcal{T}$ (a pictorial representation in which all the branch lines lie within the plane), each strip x_i will have $\tau(x_i)$ half-twists for some signed integer $\tau(x_i)$. Partition the strips $\{x_i\}$ according to those which are orientation preserving ($\tau(x_i)$ even) and those which are orientation reversing ($\tau(x_i)$ odd). Note that this partition depends on the choice of planar representation, and, in practice, one wants to choose as simple a presentation as possible. Given points $\mathbf{a}$ and $\mathbf{b}$ in Σ_{ℓ_j}, define J as in Equation (2.19), and consider the parity $\Xi \in \{0,1\}$ which keeps track of orientation

$$\Xi \equiv \left(\sum_{i=0}^{J-1} \tau(a_i) \right) \bmod 2. \tag{2.20}$$

Then define the ordering $\lhd$ on Σ_{ℓ_j} in terms of the provisional ordering $\tilde{\lhd}$ by

$$\Xi = 0 \quad : \quad \mathbf{a} \lhd \mathbf{b} \Leftrightarrow \mathbf{a}\,\tilde{\lhd}\,\mathbf{b}$$
$$\Xi = 1 \quad : \quad \mathbf{b} \lhd \mathbf{a} \Leftrightarrow \mathbf{a}\,\tilde{\lhd}\,\mathbf{b}.$$

This ordering $\lhd$ reflects the "physical" ordering of orbits on the nonorientable template $\mathcal{T}$. It is clear that this procedure can be easily implemented on a computer.

Equipped with the ordering $\lhd$, we can treat $\Sigma_{\mathcal{T}}$ as being embedded in a finite disjoint union of one-dimensional segments (although Σ_T is really a Cantor set). As such, we will introduce some notation for branch segments. Recall from Definition 2.4.2 that $\Sigma_{\mathcal{T}}$ partitions into N branch segments, where $\beta_i(\mathcal{T})$ denotes all itineraries beginning with x_i. Since this geometrically represents all orbits which begin at the x_i-strip, we will consider $\beta_i(\mathcal{T})$ as a closed interval, reflecting the total ordering $\lhd$:

Definition 2.4.4 Given $\mathcal{T}$ a template with strips $\{x_i\}_1^N$ and branch set $\beta(\mathcal{T})$, let the *ith-left-boundary*, $\partial_i^{\ell}(\mathcal{T})$, be the point of $\beta_i(\mathcal{T})$ which is $\lhd$-minimal. Similarly, let the *ith-right-boundary*, $\partial_i^{r}(\mathcal{T})$, be the point of $\beta_i(\mathcal{T})$ which is $\lhd$-maximal. The *boundary set*, $\partial(\mathcal{T})$, is given as the union of $\{\partial_i^{\ell}(\mathcal{T}), \partial_i^{r}(\mathcal{T})\}$ over i.

It is clear that $\partial(\mathcal{T})$ consists of the $2N$ eventually periodic orbits which together comprise the boundary of the template.

Remark 2.4.5 In flows whose templates have a single branch line, corresponding to a global cross section, it is natural to identify the period of a closed orbit with the number of intersections with the branch line. Often, this coincides with the number of strands in a closed braid representation. In the more general context of the present work, we identify the period of an orbit with the number of intersections of the orbit with all branch lines (hence, the period of the orbit for the return map induced by the branch lines). In all cases it coincides with the length of the periodically repeating block in the corresponding orbit word. We will thus sometimes refer to this block length as the *symbolic period*.

For a given template $\mathcal{T}$, the symbolic data; $\Sigma_\mathcal{T}, A_\mathcal{T}, \beta(\mathcal{T}), \partial(\mathcal{T})$, and $\lhd$, encode the dynamics and the combinatorial structure of the template. They do not, however, specify the topology of the enclosed orbits, nor do they provide invariants of the underlying link $L_\mathcal{T}$, since one may change the embedding of $\mathcal{T}$ without altering the symbolic data. Conversely, we may refine the Markov partition (i.e., increase the number of branch segments) without discarding any orbits from the template: see Figure 2.19 for an example. Even so, these symbolic tools do become useful in describing proper infinite sublinks and in describing the relative placement of complicated orbits.

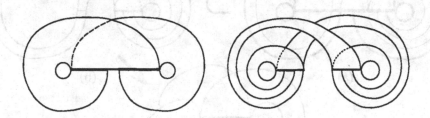

Figure 2.19: Two templates which carry the same dynamics and topology on the periodic orbits, but which have different symbolic structures.

2.4.2 Subtemplates and template inflations

In the study of templates and their properties, there are varying "scales" at which one may choose to work. Often, the knowledge of which types of individual knots or links appear on a given template is useful: this is a "small scale" question. For example, in §4.2, we will see how careful bounds on the genus of individual horseshoe knots can be used to derive uniqueness and bifurcation results in a family of Hénon maps. On the other hand, one might ask "large scale" questions about whether two entire templates (including *all* their orbits) are equivalent. This perspective will come into play in Chapter 5. Here, however, we focus on a "medium" scale question: we examine subsets of orbits which are proper yet non-finite. These are described via the notion of *subtemplates*.

Definitions and examples

Definition 2.4.6 A *subtemplate* S of a template T, written $S \subset T$, is a topological subset of T which, equipped with the restriction of the semiflow of T to S, satisfies the definition of a template (Definition 2.2.1).

A subtemplate is thus a compact branched submanifold with boundary, for which the original semiflow restricts to an expanding semiflow.

Example 2.4.7 An example of a subtemplate of the Lorenz template is given in Figure 2.20. When we "cut" along the boundaries of the subtemplate $S \subset \mathcal{L}(0,0)$, we can remove S and isotope it into the nice presentation of Figure 2.20(c). The move from part (b) to part (c) is one that we will encounter often in the remainder of this work: it is the so-called *belt trick*, in which a curl is exchanged for a full twist.

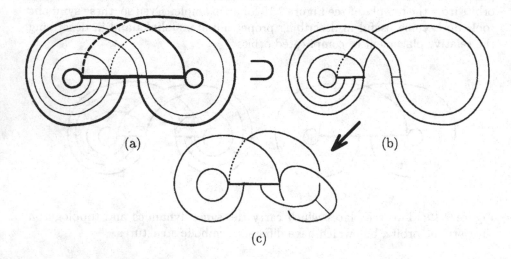

Figure 2.20: (a) a subtemplate S within $\mathcal{L}(0,0)$, (b) when removed from $\mathcal{L}(0,0)$, (c) is isotopic to $\mathcal{L}(0,2)$.

Note that S is a very special subtemplate of $\mathcal{L}(0,0)$ in that S is diffeomorphic to $\mathcal{L}(0,0)$ (it is in fact isotopic to $\mathcal{L}(0,2)$ — recall Figure 2.8(a)). Although this is not always the case, a diffeomorphic relationship between a template and a subtemplate opens up a new set of objects.

Definition 2.4.8 A *template renormalization* of a template T is a smooth embedding $\mathfrak{R} : T \hookrightarrow T$ which respects orbits (*i.e.*, it commutes with the semiflow).

It follows from Definition 2.4.6 that the image of a template renormalization $\mathfrak{R}(\mathcal{T})$ is a subtemplate of $\mathcal{T}$ which is diffeomorphic to $\mathcal{T}$. Returning to Example 2.4.7, the subtemplate $\mathcal{S} \subset \mathcal{L}(0,0)$ is the image of a template renormalization $\mathfrak{R} : \mathcal{L}(0,0) \hookrightarrow \mathcal{L}(0,0)$.

The terminology for Definition 2.4.8 arises from the one-dimensional return maps for a template induced by the branch lines [47]. The image of a template renormalization is merely a renormalization of the return maps, suspended in accordance with the template structure. We prefer, however, to think in terms of renormalizing the branched two-manifold itself, since template renormalizations carry with them the topology of the periodic orbits as well.

Since a template renormalization $\mathfrak{R}$ acts on orbits of $\mathcal{T}$ diffeomorphically, $\mathfrak{R}$ maps periodic orbits to periodic orbits: hence, there is a topological action on the underlying link $L_{\mathcal{T}}$. When this action is trivial, we say that the renormalization is *isotopic*.

Definition 2.4.9 Let $\mathfrak{R} : \mathcal{T} \hookrightarrow \mathcal{T}$ be a renormalization on an embedded template $\mathcal{T} \subset S^3$ and let $i_{\mathcal{T}}$ denote the inclusion of $\mathcal{T}$ into S^3. If $i_{\mathcal{T}}$ and $i_{\mathcal{T}} \circ \mathfrak{R}$ are isotopic embeddings of $\mathcal{T}$ in S^3, then $\mathfrak{R}$ is an *isotopic renormalization*.

The existence of a template renormalization immediately allows one to iterate $\mathfrak{R}$ on the renormalized subtemplate. This procedure enables one to extract very "deep" subtemplates, which may contain significant information about the periodic orbit link. When the renormalization has trivial action on the topology of the underlying periodic orbit link, we may iterate to obtain complicated subtemplates whose orbits have extremely long symbolic period, while still controlling the individual knot and link types.

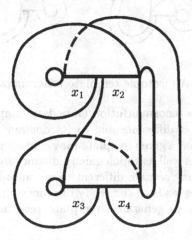

Figure 2.21: The template $\mathcal{V}$.

Example 2.4.10 The first example of an isotopic template renormalization (without that terminology) was given by M. Sullivan [169]. Let $\mathcal{V}$ denote the embedded template of Figure 2.21, having two branch lines with a total of four strips, $\{x_1, x_2, x_3, x_4\}$. The template $\mathcal{V}$ is embedded such that none of its strips are knotted or twisted, but note that it contains crossings of both positive and negative types. The renormalization taking $\mathcal{V}$ into itself is illustrated in Figure 2.22, from which it is clear that the image is isotopic to the domain, for the positive and negative twists produced by the belt trick exactly cancel.

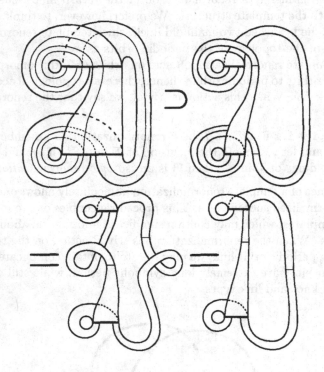

Figure 2.22: An isotopic template renormalization on $\mathcal{V}$.

Thus far, a template renormalization embeds a template within itself, and any subtemplate which is diffeomorphic to its domain can be described by a renormalization. However, a given template may contain numerous subtemplates which are dynamically as well as topologically distinct from the original, just as one-dimensional maps may contain different maps embedded deep within. This phenomenon in 1-d maps leads to the study of renormalizations between classes of maps [73]. We wish to generalize template renormalizations in a similar manner.

Definition 2.4.11 A *template inflation* is a smooth embedding $\mathfrak{R} : \mathcal{S} \hookrightarrow \mathcal{T}$ of a template $\mathcal{S}$ into a template $\mathcal{T}$ which respects orbits (*i.e.*, it commutes with the semiflow).

It follows from Definition 2.4.6 that the image of a template inflation $\mathfrak{R}(\mathcal{S})$ is a subtemplate of $\mathcal{T}$. A template renormalization is a special form of a template inflation, and we will often use the more general term. The analogous notion of an isotopic template inflation follows:

Definition 2.4.12 Let $\mathfrak{R} : S \hookrightarrow \mathcal{T}$ be an inflation of a template $S \subset S^3$ into a template $\mathcal{T} \subset S^3$. Let i_S and $i_\mathcal{T}$ denote inclusion of S and $\mathcal{T}$ respectively into S^3. If i_S and $i_\mathcal{T} \circ \mathfrak{R}$ are isotopic embeddings of S in S^3, then $\mathfrak{R}$ is an *isotopic inflation*.

There are many basic questions about subtemplates and template inflations, *e.g.*:

Problem 2.4.13 Given a template $\mathcal{T}$, which templates embed [abstractly] in $\mathcal{T}$ (*i.e.*, which are images of inflations)? Given an embedded template $\mathcal{T}$, what are all the subtemplates of $\mathcal{T}$ (*i.e.*, which are images of isotopic inflations)?

We will obtain in §3.3 the surprising answer that *all* orientable templates embed in any $\mathcal{T}$ (after a slight perturbation at the branch lines). Furthermore, we will show that certain templates contain isotopic copies of *all* templates as subtemplates.

The goal of working with template inflations is to understand properties of deep, complicated subtemplates within a given template. To that end, isotopic inflations are useful, in that we can keep track of the knots and links which live "deep within" a template by pulling back the isotopy. To keep track of where exactly these complicated subtemplates lie, we use the induced action of an inflation on the itinerary space in order to derive "coordinates" for a subtemplate associated to a given inflation.

Symbolic actions of inflations

Lemma 2.4.14 *A template inflation $\mathfrak{R} : S \hookrightarrow \mathcal{T}$ induces an embedding $\mathfrak{R} : \Sigma_S \hookrightarrow \Sigma_\mathcal{T}$ whose action is to inflate each symbol $\{x_i : i = 1..M\}$ of Σ_S to a finite admissible word $\{\mathbf{w}^i = w_1 \ldots w_{n(i)} : i = 1..N\}$ in the symbols of $\Sigma_\mathcal{T}$.*

Proof: by Definition 2.4.11, $\mathfrak{R}$ maps the branch lines of S into branch lines of $\mathcal{T}$. Hence, each strip of S (corresponding to a generator x_i of Σ_S) is mapped to a finite sequence of strips in $\mathcal{T}$, corresponding to a finite admissible itinerary for $\mathcal{T}$. □

The image under $\mathfrak{R}$ of any orbit on S is thus obtained by "inflating" each symbol x_i in the itinerary by the word $\mathbf{w}^i$ (which in some cases may consist of a single letter). This immediately implies the following useful result:

Corollary 2.4.15 *Given $\mathfrak{R} : S \hookrightarrow \mathcal{T}$ a template inflation, the branch set and the boundary of the subtemplate $\mathfrak{R}(S)$ are given by*

$$
\begin{aligned}
\beta_i(\mathfrak{R}(S)) &= \mathfrak{R}(\beta_i(S)) &= \{\mathfrak{R}(a); a \in \beta_i(S)\} \\
\partial(\mathfrak{R}(S)) &= \mathfrak{R}(\partial(S)) &= \{\mathfrak{R}(a); a \in \partial(S)\}.
\end{aligned}
\tag{2.21}
$$

We wish to consider the branch set $\beta(\mathfrak{R}(\mathcal{S}))$ as a set of "coordinates" consisting of N "subintervals" of the branch set of $\mathcal{T}$ which indicate where $\mathcal{S}$ resides within $\mathcal{T}$. We note that the image of a branch segment under an inflation is not an interval in the sense that all orbits between its endpoints are not necessarily part of the subtemplate (recall there are "gaps" in the branch lines). Yet, if we consider the N subintervals given by $\beta(\mathfrak{R}(\mathcal{S}))$, we have a relative measure of the *depth* of an inflation. For example, if a template $\mathcal{T}$ contains a nested sequence of subtemplates $\mathcal{T}_n \subset \ldots \subset \mathcal{T}_2 \subset \mathcal{T}_1 \subset \mathcal{T}$, then the same inclusion exists on the branch sets $\beta(\mathcal{T}_i)$ within $\beta(\mathcal{T})$. Or, given two subtemplates of $\mathcal{T}$, the information encoded in their symbolic branch sets can be used to determine whether these subtemplates are disjoint, or which subtemplate is "closer" (under $\lhd$) to a given periodic orbit.

Example 2.4.16 For an example which will demonstrate the symbolic actions of an isotopic inflation, we return to the isotopic renormalization of $\mathcal{V}$ from Example 2.4.10. From Figure 2.22, one traces the image of the four strips $\{x_1, x_2, x_3, x_4\}$ to obtain the symbolic action:

$$\mathfrak{D} : \mathcal{V} \hookrightarrow \mathcal{V} \quad \begin{cases} x_1 \mapsto x_1 \\ x_2 \mapsto x_1 x_2 \\ x_3 \mapsto x_3 \\ x_4 \mapsto x_3 x_4 \end{cases}. \tag{2.22}$$

The branch segments of the subtemplate are given by

$$\begin{aligned} \beta_1(\mathfrak{D}(\mathcal{V})) &= \mathfrak{D}\left([(x_1)^\infty, x_1(x_2 x_4)^\infty]\right) = [(x_1)^\infty, x_1(x_1 x_2 x_3 x_4)^\infty] \\ \beta_2(\mathfrak{D}(\mathcal{V})) &= \mathfrak{D}\left([x_2(x_3)^\infty, (x_2 x_4)^\infty]\right) = [x_1 x_2(x_3)^\infty, (x_1 x_2 x_3 x_4)^\infty] \\ \beta_3(\mathfrak{D}(\mathcal{V})) &= \mathfrak{D}\left([(x_3)^\infty, x_3(x_4 x_2)^\infty]\right) = [(x_3)^\infty, x_3(x_3 x_4 x_1 x_2)^\infty] \quad (2.23) \\ \beta_4(\mathfrak{D}(\mathcal{V})) &= \mathfrak{D}\left([x_4(x_1)^\infty, (x_4 x_2)^\infty]\right) = [x_3 x_4(x_1)^\infty, (x_3 x_4 x_1 x_2)^\infty]. \end{aligned}$$

The boundary components of the subtemplate, $\partial(\mathfrak{D}(\mathcal{V}))$, are given by the endpoints of the intervals above.

We encourage the reader to work through this example carefully, correlating the geometric description of Figure 2.22 with the symbolic description of Equation (2.22). This procedure is used extensively in Chapter 3.

Unfortunately, one cannot endow the symbolic structure with very much information about the topology of the infinite link. However, the hyperbolicity of the underlying flow does give a nice structure to the space $\Sigma_\mathcal{T}$ which we hope to utilize as much as possible. By looking at the ordering $\lhd$ and by considering the relationship between iterated subtemplates and their "coordinates" in terms of branch sets, we have a set of tools for describing and manipulating "deep" sublinks of the link of periodic orbits. We will use these in the next chapter to prove some basic, as well as some surprising, results.

Chapter 3: Template Theory

In this chapter, we use the tools of Chapter 2 to build a collection of general results on templates and template links, noting applications to the dynamics of three-dimensional flows along the way. We begin in §3.1 with a treatment of properties of the individual knots and links which are supported on a given embedded template. Then, in §3.2, we use the methods developed in §3.1 and the previous chapter to prove the existence (and abundance) of *universal templates*: templates which contain *all* knots and links among their closed orbits. In §3.3, we continue this line of inquiry to examine the *subtemplate problem:* the enumeration of all subtemplates of a given embedded template.

These results, which are fairly general in nature, will lead to numerous specific conclusions in this and in subsequent chapters when applied to the examples introduced in §2.3.

3.1   Knotted orbits on templates

Question 1 *Given an embedded template T, does it contain a nontrivial knot? How many such knots are present? How are these distributed?*

In this section, we will answer Question 1, giving applications to the dynamics of flows.

3.1.1   Alexander's Theorem for templates

In many of the results to follow, we will need to represent template knots and links as closed braids. We begin with an analogue of braiding for templates:

Definition 3.1.1 A template T is said to be *braided* if T is embedded in $D^2 \times S^1$ in such a way that every closed orbit on T is a closed braid: that is, each meridional disc $D^2 \times \{\theta\}$ intersects the curve transversely in a fixed number of points. A template is said to be *positive* if it can be braided in such a way that every closed orbit is a closed positive braid.

Recall Alexander's Theorem (Theorem 1.1.13), which states that any link is isotopic to a closed braid. The corresponding statement for templates is also true, as shown by Franks and Williams [58].

Theorem 3.1.2 (The Alexander Template Theorem: Franks and Williams [58]) *Any template T may be isotoped so that it is a closed braided template. Furthermore, if T is orientable, it may be arranged such that in a planar projection, all the strips of T are flat (untwisted).*

The proof closely follows that of Alexander's Theorem for links [3]: a nice account of the latter can be found in [33, Prop. 2.14]. In the proof of Alexander's Theorem, one chooses a tentative braid axis, and then iteratively "flips" strands of the link about the braid axis until they are all aligned. Here, instead of wrapping strands about a braid axis, one manipulates strips. To obtain a flat presentation, one uses the belt trick of Example 2.4.7 to exchange a full twist for an additional trip about the braid axis. Half twists, which arise in non-orientable templates, of course cannot be straightened.

3.1.2   Concatenation of template knots

Given two periodic points of Σ_{ℓ_j} – the set of all orbits starting on the branch line ℓ_j – we wish to define an "addition" operation which has both symbolic and topological interpretations.

Definition 3.1.3 Let $\mathbf{a}^\infty$ and $\mathbf{b}^\infty$ be distinct periodic points of Σ_{ℓ_j}. Then the *concatenation* of $\mathbf{a}^\infty$ and $\mathbf{b}^\infty$, denoted $\mathbf{a}^\infty \oplus \mathbf{b}^\infty$, is the point $(\mathbf{ab})^\infty \in \Sigma_{\ell_j}$.

Remark 3.1.4 The concatenation operation is well-defined: since $\mathbf{a}^\infty$ and $\mathbf{b}^\infty$ are both points on a particular branch line ℓ_j, the orbit $(\mathbf{ab})^\infty$ must be admissible. Note, however, that $\mathbf{ab}$ may equal $\mathbf{u}^k$ for $k > 1$ and some $\mathbf{u}$, as in $x_1^2 x_2 x_1$ concatenated with $x_1 x_2$. In this case, we would say $\left(x_1^2 x_2 x_1\right)^\infty \oplus \left(x_1 x_2\right)^\infty = \left(x_1^2 x_2\right)^\infty$.

Given the concatenation operation, we wish to understand the topological action on periodic orbits. We begin with a class of concatenations which behave nicely.

Definition 3.1.5 Choose two distinct points $\mathbf{u}$ and $\mathbf{v} \in \Sigma_{\ell_j}$ and assume that $\mathbf{u} \lhd \mathbf{v}$. Define $(\mathbf{u}, \mathbf{v})$ to be the set of all point $\mathbf{x} \in \Sigma_{\ell_j}$ such that $\mathbf{u} \lhd \mathbf{x} \lhd \mathbf{v}$. Then $\mathbf{u}$ and $\mathbf{v}$ are said to be *adjacent* if,

$$\left\{\sigma^k \mathbf{u}\right\}_{k>0} \cap (\mathbf{u}, \mathbf{v}) = \left\{\sigma^k \mathbf{v}\right\}_{k>0} \cap (\mathbf{u}, \mathbf{v}) = \emptyset. \tag{3.1}$$

Thus, $\mathbf{u}$ and $\mathbf{v}$ are adjacent if no other points on their orbits appear between $\mathbf{u}$ and $\mathbf{v}$.

In order to simplify the next few results, we circumvent the exceptional cases of Remark 3.1.4:

Lemma 3.1.6 *If $\mathbf{a}$ and $\mathbf{b}$ are distinct nontrivial words and $\mathbf{ab} = \mathbf{u}^k$ for $k > 1$ and some $\mathbf{u}$, then $\mathbf{a}^\infty$ and $\mathbf{b}^\infty$ are not adjacent.*

Proof: Decompose $\mathbf{a} = \mathbf{u}^i \mathbf{a}'$ and $\mathbf{b} = \mathbf{b}' \mathbf{u}^j$, where $i + j = k - 1$ and $\mathbf{a}'\mathbf{b}' = \mathbf{u}$. Assuming (arbitrarily) that $\mathbf{a} \lhd \mathbf{b}$ and that $i > 0$, consider the point $\left(\mathbf{u}^{i-1}\mathbf{a}'\mathbf{u}\right)^\infty$, which is a shift of $\mathbf{a}^\infty$. Then, since $\mathbf{a}^\infty \lhd \mathbf{u}^\infty \lhd \mathbf{b}^\infty$, it follows that

$$\mathbf{a}^\infty \lhd \left(\mathbf{u}^{i-1}\mathbf{a}'\mathbf{u}\right)^\infty \lhd \mathbf{u}^\infty \lhd \mathbf{b}^\infty, \tag{3.2}$$

whence it follows that $\mathbf{a}^\infty$ and $\mathbf{b}^\infty$ are not adjacent.   □

The concatenation of adjacent orbits is similar in spirit to taking a connected sum: only one crossing is added.

Lemma 3.1.7 *Let $\mathcal{T}$ be an embedded template, and let $\mathbf{a}^\infty$ and $\mathbf{b}^\infty$ be adjacent periodic points in Σ_{ℓ_j}. The planar presentation of the knot corresponding to $\mathbf{a}^\infty \oplus \mathbf{b}^\infty$ differs from that of the link corresponding to $\mathbf{a}^\infty$ union $\mathbf{b}^\infty$ by the addition of a single crossing (as illustrated in Figure 3.1).*

Proof: Place $\mathcal{T}$ in a planar presentation and consider the branch line ℓ_j which contains the points $\mathbf{a}^\infty \lhd \mathbf{b}^\infty$. By isotoping $\mathcal{T}$ if necessary, a neighborhood of ℓ_j will appear locally as in Figure 3.1(a) – there are two cases depending on which strip is "on top." By properties of the ordering $\lhd$, it follows that

$$\mathbf{a}^\infty \lhd (\mathbf{ab})^\infty \lhd (\mathbf{ba})^\infty \lhd \mathbf{b}^\infty, \tag{3.3}$$

so that the concatenated orbit appears as in Figure 3.1(b): there is a new crossing whose sign is dependent upon the stacking order of strips. The orbit $(\mathbf{ab})^\infty$ follows $\mathbf{a}$ then $\mathbf{b}$: the ordering of points on other branch lines does not change. More specifically, if, on any branch line, $(\sigma^i \mathbf{a})^\infty \lhd (\sigma^j \mathbf{a})^\infty$, then it follows that $(\sigma^i(\mathbf{ab}))^\infty \lhd (\sigma^j(\mathbf{ab}))^\infty$ for any $i, j < |\mathbf{a}|$. Hence, $\mathbf{a}^\infty \oplus \mathbf{b}^\infty$ may be isotoped to the link $\mathbf{a}^\infty$ union $\mathbf{b}^\infty$ with a single crossing inserted at the branch line as specified. □

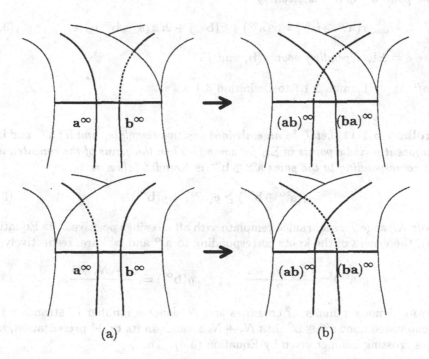

Figure 3.1: Concatenation of adjacent periodic points effects a local change as above.

Lemma 3.1.7 immediately yields:

Corollary 3.1.8 *Let $\mathcal{T}$ be an embedded template, and let $\mathbf{a}^\infty$ and $\mathbf{b}^\infty$ be adjacent periodic points in Σ_{ℓ_j} with self-crossing numbers c_a and c_b respectively. Then, the self-crossing number of the concatenation $\mathbf{a}^\infty \oplus \mathbf{b}^\infty$ is given by*

$$c_{a \oplus b} = c_a + c_b + 2\ell k\,(\mathbf{a}^\infty, \mathbf{b}^\infty) + \epsilon, \tag{3.4}$$

where $\epsilon = \pm 1$, depending upon $\mathbf{a}^\infty, \mathbf{b}^\infty$, and $\mathcal{T}$, is the sign of the crossing of Lemma 3.1.7.

Definition 3.1.9 The *twist* of a ribbon (annulus or Möbius strip) in S^3 with c crossings and t signed half-twists (in a given planar presentation) is given as $c + \frac{1}{2}t$ and is an isotopy invariant (see Lemma 5.3.4 for a proof). Given K a closed orbit on a template $\mathcal{T}$, the *twist* of K, τ_K, is defined to be the twist of the normal bundle of $\mathcal{T}$ restricted to K. That is, the bundle of normal directions to $\mathcal{T}$ along K is an embedded *ribbon* in S^3 with twist τ_K. Equivalently, this ribbon is the local stable manifold to the orbit.

Corollary 3.1.10 *Let $\mathcal{T}$ be an embedded template, and let $\mathbf{a}^\infty$ and $\mathbf{b}^\infty$ be adjacent periodic points in Σ_{ℓ_j}. Then the twist of the concatenated knot corresponding to the point $\mathbf{a}^\infty \oplus \mathbf{b}^\infty$ is given by*

$$\tau(\mathbf{a}^\infty \oplus \mathbf{b}^\infty) = \tau(\mathbf{a}^\infty) + \tau(\mathbf{b}^\infty) + 2\ell k\,(\mathbf{a}^\infty, \mathbf{b}^\infty) + \epsilon, \tag{3.5}$$

where $\epsilon = \pm 1$, depending upon $\mathbf{a}, \mathbf{b}$, and $\mathcal{T}$.

Proof: Apply Lemma 3.1.7 to Definition 3.1.9.   □

Corollary 3.1.11 *Let $\mathcal{T}$ be an embedded positive template, and let $\mathbf{a}^\infty$ and $\mathbf{b}^\infty$ be adjacent periodic points in Σ_{ℓ_j} for some j. Then the genus of the concatenated knot corresponding to the point $\mathbf{a}^\infty \oplus \mathbf{b}^\infty$ is bounded below as*

$$g(\mathbf{a}^\infty \oplus \mathbf{b}^\infty) \geq g(\mathbf{a}^\infty) + g(\mathbf{b}^\infty). \tag{3.6}$$

Proof: Arrange $\mathcal{T}$ as a braided template with all crossings positive. Via Equation (1.3), the genera of the knots corresponding to $\mathbf{a}^\infty$ and $\mathbf{b}^\infty$ are, respectively,

$$g(\mathbf{a}^\infty) = \frac{c_a - N_a + 1}{2} \quad ; \quad g(\mathbf{b}^\infty) = \frac{c_b - N_b + 1}{2}, \tag{3.7}$$

where c denotes number of crossings and N denotes number of strands. The concatenated knot $\mathbf{a}^\infty \oplus \mathbf{b}^\infty$ has $N_a + N_b$ strands in its braid presentation, and it has crossing number given by Equation (3.4). Thus,

$$
\begin{aligned}
g(\mathbf{a}^\infty \oplus \mathbf{b}^\infty) &= \frac{c_a + c_b + 2\ell k\,(\mathbf{a}^\infty, \mathbf{b}^\infty) + \epsilon - (N_a + N_b) + 1}{2} \\
&= g(\mathbf{a}^\infty) + g(\mathbf{b}^\infty) + \frac{2\ell k\,(\mathbf{a}^\infty, \mathbf{b}^\infty) - 1 + \epsilon}{2}. \tag{3.8}
\end{aligned}
$$

Since all crossings are positive prior to and after concatenation, $\ell k\,(\mathbf{a}^\infty,\mathbf{b}^\infty) \geq 0$. If $\epsilon = -1$, then in concatenation we have *removed* a (positive) crossing; thus, for $\epsilon = -1$, $\ell k\,(\mathbf{a}^\infty,\mathbf{b}^\infty) > 0$ prior to concatenation, and the result follows. For $\epsilon = +1$, it is obviously true. □

Corollary 3.1.11 gives a partial answer to a generalization of a conjecture of Williams's:

Conjecture 3.1.12 Let $\mathcal{T}$ be a positive embedded template. Let $\mathbf{a}^\infty$ and $\mathbf{b}^\infty$ be periodic itineraries in Σ_{ℓ_j} (not necessarily adjacent). Then, genus is monotonic under the $\oplus$ operation:[1]

$$g((\mathbf{ab})^\infty) \geq g(\mathbf{a}^\infty) + g(\mathbf{b}^\infty). \qquad (3.9)$$

We will use the $\oplus$ operation in the next subsection, when we describe where on a template knots live.

3.1.3 The existence of knots on a template

Theorem 3.1.13 *Given an embedded template $\mathcal{T}$, there exists a nontrivial knot as an orbit on $\mathcal{T}$.*

Proof: Our proof is in the spirit of Proposition 4.4 of [58], in that we rely upon the Bennequin inequality.[2] Arrange $\mathcal{T}$ as a braided template as per Theorem 3.1.2. Choose $\mathbf{a}^\infty$ and $\mathbf{b}^\infty$ in some branch set component Σ_{ℓ_j} with $\mathbf{a}^\infty$ and $\mathbf{b}^\infty$ adjacent. Assume that the twist of $\mathbf{a}^\infty$ or $\mathbf{b}^\infty$ is nonzero. If not, then replace $\mathbf{a}^\infty$ with $\mathbf{a}^\infty \oplus \mathbf{b}^\infty$. By Corollary 3.1.10, the twist of the concatenated knot is nonzero and this orbit is still adjacent to $\mathbf{b}^\infty$.

Given $\mathbf{a}^\infty$ and $\mathbf{b}^\infty$ with $\tau(\mathbf{a}^\infty) \neq 0$, concatenate repeatedly to form the orbit

$$(\mathbf{a}^n\mathbf{b})^\infty = \mathbf{a}^\infty \oplus (\mathbf{a}^\infty \oplus (\cdots(\mathbf{a}^\infty \oplus \mathbf{b}^\infty)\cdots)). \qquad (3.10)$$

We will use the Bennequin inequality, Equation (1.5), to bound the genus of this knot. By Corollary 3.1.8, the self-crossing number of $(\mathbf{a}^n\mathbf{b})^\infty$ is

$$c_{a^n b} = c_a n^2 + \frac{1}{2}t_a n(n-1) + c_b + (2\ell k\,(\mathbf{a}^\infty,\mathbf{b}^\infty) + \epsilon)\,n, \qquad (3.11)$$

where c_a (resp. c_b) is the self-crossing number of $\mathbf{a}^\infty$ (resp. $\mathbf{b}^\infty$), t_a is the signed number of half-twists in the presentation of the embedded normal bundle of $\mathbf{a}^\infty$, and $\epsilon = \pm 1$. See Figure 3.2 for the count of the terms quadratic in n. By Equation (1.5),

$$2g((\mathbf{a}^n\mathbf{b})^\infty) \geq \begin{vmatrix} c_a n^2 + \frac{1}{2}t_a n(n-1) + c_b + (2\ell k\,(\mathbf{a}^\infty,\mathbf{b}^\infty) + \epsilon)\,n \\ -(nN_a + N_b) + 1 \end{vmatrix}, \qquad (3.12)$$

---

[1] An exception occurs as in Remark 3.1.4, which we could circumvent by defining the genus of $\left(\mathbf{u}^k\right)^\infty$ to be k times the genus of $\mathbf{u}^\infty$.

[2] It is an open (and challenging) problem to prove this theorem without resorting to Bennequin's inequality.

Since the twist $\tau_a \neq 0$, $c_a + \frac{1}{2}t_a \neq 0$; hence, $c_{a^n b}$ is quadratic in n as per (3.11). Thus, for some n, the genus of $(a^n b)^\infty$ is nonzero.   □

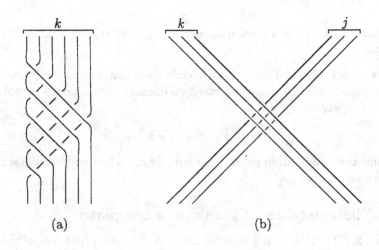

(a)   (b)

Figure 3.2: (a) each half-twist on k-strands yields $\frac{1}{2}k(k-1)$ crossings; (b) each crossing of k-strands over j-strands yields kj crossings.

Corollary 3.1.14 *Given an embedded template $\mathcal{T}$, there exists an infinite number of distinct knot types as orbits on $\mathcal{T}$.*

Proof: Let $n \to \infty$ above.   □

From this, we may recover the Franks-Williams Theorem for flows on S^3:

Theorem 3.1.15 (Franks and Williams [58]) *Any C^2-flow on S^3 which has positive topological entropy must display an infinite number of distinct knot types as closed orbits.*

Proof: By a [deep] theorem of Katok [97], a C^2 flow with positive topological entropy must contain a hyperbolic periodic orbit which has a transverse homoclinic connection. The Poincaré-Birkhoff-Smale Theorem, Theorem 1.2.33, then asserts the existence of an embedded Smale horseshoe in the flow. By the Template Theorem, this basic set collapses to an embedded template in S^3 which captures knot and link types. This template, and hence the flow, supports an infinite number of knot types by Corollary 3.1.14.   □

Remark 3.1.16 Theorem 3.1.15 is a beautiful result, yielding a great deal of topological information from purely dynamical data. The connection is thus established: dynamically complicated hyperbolic flows on S^3 force topologically complicated knots as orbits. Several converses exist: for an example, see the

Morgan-Wada Theorem in Appendix A. Another well-known converse is the Seifert Conjecture, recently resolved in the smooth case by K. Kuperberg [107]. This result states that there exist smooth nonsingular flows on S^3 containing *no* periodic orbits whatsoever.

From Theorem 3.1.13 we may also derive information about how knots are distributed on Σ_T. We show that the nontrivial knots do not confine themselves to any proper subregion.

Corollary 3.1.17 *Let T be an* irreducible *template — that is, the subshift of finite type defined on Σ_T has a dense orbit. Then, given any point $\mathbf{x}$ in Σ_T, there exists an infinite number of distinct knot types represented in an arbitrarily small neighborhood of $\mathbf{x}$.*

Proof: Choose a small ϵ-neighborhood N_ϵ of $\mathbf{x}$ in Σ_T and pick two distinct periodic points $\mathbf{a}^\infty$ and $\mathbf{b}^\infty \in N_\epsilon$ (this is always possible since the periodic points are dense in Σ_T for T irreducible). If necessary, shift $\mathbf{b}^\infty$ to be adjacent to $\mathbf{a}^\infty$ — this does not remove it from N_ϵ. Consider the template inflation

$$\mathfrak{R} : \mathcal{L}(\tau_a, \tau_b) \hookrightarrow T \qquad \left\{ \begin{array}{l} x_1 \mapsto \mathbf{a} \\ x_2 \mapsto \mathbf{b} \end{array} \right. , \tag{3.13}$$

where τ_a (τ_b resp.) is the twist of $\mathbf{a}^\infty$ ($\mathbf{b}^\infty$ resp.) and $\mathcal{L}(m,n)$ is the Lorenz-like template of type (m,n) (see §2.3.1). This inflation is well-defined since $\mathbf{a}^\infty$ and $\mathbf{b}^\infty$ are adjacent. The image of $\mathfrak{R}$ has branch set

$$\beta\{\mathfrak{R}(\mathcal{L}(\tau_a, \tau_b))\} = \left\{ \begin{array}{ll} [\mathbf{a}^\infty, \mathbf{b}^\infty] & : \tau_a, \tau_b \text{ even} \\ [\mathbf{a}^\infty, \mathbf{ba}^\infty] & : \tau_a \text{ even}, \tau_b \text{ odd} \\ [\mathbf{ab}^\infty, \mathbf{ba}^\infty] & : \tau_a, \tau_b \text{ odd} \end{array} \right. , \tag{3.14}$$

which is contained within a 2ϵ-neighborhood of $\mathbf{a}^\infty$. By Corollary 3.1.14, this subtemplate contains an infinite set of distinct knot types. □

Remark 3.1.18 Any template obtained from a basic set of a flow is irreducible, since basic sets have dense orbits. A non-irreducible template is, from our perspective, an anomaly.

3.1.4 Accumulations of knots

Knowing that knot types are "densely packed" on any given template says nothing about their precise distribution. What are the chances of a figure-eight knot living arbitrarily close to a trefoil? To an unknot? To answer this (in part), we will explore the special role played by unknots with zero twist.

Proposition 3.1.19 *Let T be an embedded template. Suppose that some point $\mathbf{u}^\infty \in \Sigma_{\ell_j}$ represents an unknotted periodic orbit with zero twist. Then, for every periodic point $\mathbf{a}^\infty$ in Σ_{ℓ_j} such that $\mathbf{a}^\infty$ and $\mathbf{u}^\infty$ are separable, there exist infinitely many periodic points in Σ_{ℓ_j} which have the same knot type as $\mathbf{a}^\infty$, and these accumulate onto $\mathbf{u}^\infty$.*

Proof: Assume (after shifting perhaps) that $\mathbf{a}^\infty$ and $\mathbf{u}^\infty$ are adjacent. We claim that the concatenation $\mathbf{u}^\infty \oplus \mathbf{a}^\infty = (\mathbf{ua})^\infty$ is the connected sum of the two original knots.

Since $\mathbf{a}^\infty$ and $\mathbf{u}^\infty$ represent separable knots, there is a 2-sphere S^2 which bounds the knots on opposite sides. By placing the sphere in general position, we may assume that S^2 intersects the template $\mathcal{T}$ transversally. Denote by I the subset of the branch line ℓ_j which is bounded by the points $\mathbf{u}^\infty$ and $\mathbf{a}^\infty$.

Let $N \subset S^3$ denote a tubular neighborhood of $\mathbf{u}^\infty \cup I \cup \mathbf{a}^\infty$ in S^3. We claim that $N \cap \mathcal{T}$ is isotopic to the configuration of Figure 3.3. To show this, note that the space $S^3 \setminus N$ is isotopic to a solid torus (the complement of the unknot $\mathbf{u}^\infty$) with an interior solid torus removed (a neighborhood of the knot $\mathbf{a}^\infty$) and a (perhaps knotted) hole connecting the boundaries of these solid tori, corresponding to the arc I. Since $\mathbf{a}^\infty$ and $\mathbf{u}^\infty$ are separable, the solid torus hole in inessential (it is contained within a ball in the solid torus). As such, one may use the "lightbulb trick" — if a lightbulb hangs from a knotted cord, the cord can be isotoped to one without a knot while fixing the light bulb — to show that N can be isotoped to the configuration of Figure 3.3 (see [154, p. 257]).

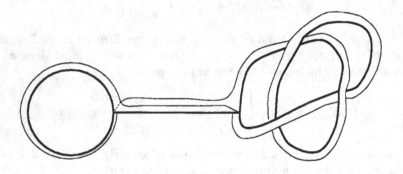

Figure 3.3: The intersection of N and $\mathcal{T}$.

Given $N \cap \mathcal{T}$ as in Figure 3.3(a), the orbit $(\mathbf{ua})^\infty$ is isotopic within $\mathcal{T}$ (hence, within S^3) to a curve within N. This isotopy involves pushing the orbit "outwards" so that it completes a circuit in a neighborhood of $\mathbf{a}^\infty$, crosses to $\mathbf{b}^\infty$ through I, continues around $\mathbf{b}^\infty$, then goes back across I.

After the isotopy, it is clear that $(\mathbf{ua})^\infty$ is the connected sum of $\mathbf{u}^\infty$ and $\mathbf{a}^\infty$. Since $\mathbf{u}^\infty$ is an unknot, $(\mathbf{ua})^\infty$ has the knot type of $\mathbf{a}^\infty$. Since $\mathbf{u}^\infty$ is unknotted and untwisted, $(\mathbf{ua})^\infty$ is also separable with respect to $\mathbf{u}^\infty$ and the process may be iterated, creating the sequence $(\mathbf{u}^k \mathbf{a})^\infty$, which accumulates on $\mathbf{u}^\infty$.   □

A converse to Proposition 3.1.19 holds for positive templates and provides a clue to the distribution of knots on templates.

Theorem 3.1.20 *Let $\mathcal{T}$ be a positive embedded template. Suppose that a sequence of distinct periodic points $\mathbf{a}_n^\infty$ in $\Sigma_\mathcal{T}$ all correspond to the same knot* *type. Then any accumulation point of this sequence of the form* $\mathbf{u}^\infty$ *represents an untwisted unknotted periodic orbit.*

Proof: Arrange $\mathcal{T}$ as a positively braided template as per Theorem 3.1.2. Given $\mathbf{u}^\infty$ an accumulation point for the sequence $\mathbf{a}_n^\infty$, reindex this latter sequence to denote the subsequence which converges to $\mathbf{u}^\infty$. For n sufficiently large, $\mathbf{a}_n^\infty$ must be of the form $\left(\mathbf{u}^k\mathbf{b}_n\right)^\infty$ for k any fixed number: this is pictured in Figure 3.4.

If $\mathbf{u}^\infty$ is nontrivially knotted, then by Equation (1.3), $c_u > N_u$, where c_u is the self-crossing number and N_u is the number of strands in the braid representation of $\mathbf{u}^\infty$. From the form of $\mathbf{a}_n^\infty = \left(\mathbf{u}^k\mathbf{b}_n\right)^\infty$, it follows that the genus of $\mathbf{a}_n^\infty$ is greater than or equal to k times the [nonzero] genus of $\mathbf{u}^\infty$. As k can be chosen arbitrarily large, the sequence $\{\mathbf{a}_n^\infty\}_n$ will not have bounded genus.

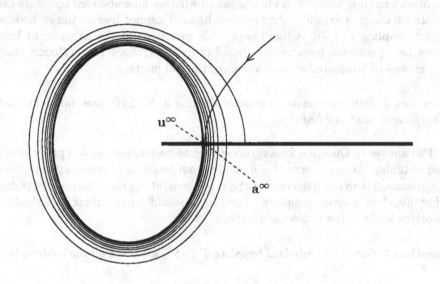

Figure 3.4: A portion of the orbit $\mathbf{a}_n^\infty$ for n large.

If $\mathbf{u}^\infty$ is an unknot of twist $\tau_u > 0$, then there are at least $\frac{1}{2}\tau_u k(k-1)$ crossings of $\mathbf{a}_n^\infty = \left(\mathbf{u}^k\mathbf{b}\right)^\infty$ with $\mathbf{u}^\infty$ (*cf.* Figure 3.2). Since, for n large, k is large, Equation (3.8) implies that the genus of the sequence $\{\mathbf{a}_n^\infty\}$ is unbounded. We conclude that $\mathbf{u}^\infty$ is an untwisted unknot. □

Theorem 3.1.20 implies that, on a positive template $\mathcal{T}$, the collection of knot types supported on $\mathcal{T}$ "accumulates" at untwisted unknots and nowhere else.

Remark 3.1.21 Let $\{\gamma_i\}_1^\infty$ be a sequence of distinct closed orbits in a flow. We say that γ_i *accumulates* on a closed orbit γ if there exists a sequence of points $\{x_i \in \gamma_i\}_1^\infty$ which have $x \in \gamma$ as an accumulation point for some $x \in \gamma$. If we consider the class of flows that have one-dimensional basic sets (*e.g.*, Smale flows) with "positive" twisting, we can lift Theorem 3.1.20 to the original flow to imply that any infinite sequence of distinct periodic orbits of bounded genus must accumulate on untwisted unknots.

Remark 3.1.22 Theorem 3.1.20 fails spectacularly for non-positive templates. Using results from the remainder of this chapter, it has recently been shown that in such cases, practically anything can occur: see Remark 3.3.12.

3.2   Universal templates

We have in Theorem 3.1.13 one extreme: every embedded template must contain a nontrivial knot, and in fact, by Corollary 3.1.14, infinitely many distinct knots. The other extreme, however, is unclear, as to whether an embedded template can contain *all* knots. Certainly, the figure-eight knot cannot live on the embedded Lorenz template $\mathcal{L}(0,0)$, as this template is positive and the figure-eight knot cannot be represented by a positive braid (recall Exercise 1.1.21). Hence, there exist classes of templates which do not contain all knots.

Question 2 *Does there exist an embedded template $\mathcal{T} \subset S^3$ containing all knots as periodic orbits? All links?*

The answer to Question 2 was conjectured to be *no* [24]: we will prove otherwise, outlining the arguments of [69], while providing a more general perspective.

Question 2 is to some degree not the most general approach to understanding "what lives" in a given template. Focusing instead on the class of embedded templates leads to the following question:

Question 3 *Given an embedded template $\mathcal{T} \subset S^3$, what are all the subtemplates of $\mathcal{T}$?*

In this section, we tackle Question 2 by using methods suited for answering Question 3.

3.2.1   Examples of subtemplate structures

Lorenz-like templates

As a basic example of a subtemplate question, recall Problem 2.3.6 concerning the relationships between the Lorenz-like templates of §2.3.1. We derive a partial answer in this subsection, following [168], but using the symbolic methods of this monograph.

In Figure 2.20 of §2.4.2, we proved that $\mathcal{L}(0,2) \subset \mathcal{L}(0,0)$ via an isotopic inflation. In the following, we use the symbolic descriptions of §2.4 to list a slightly more complete collection of isotopic inflations relating these templates.

Proposition 3.2.1 *The following template inflations act isotopically:*

$$\mathcal{L}(0, n+2) \hookrightarrow \mathcal{L}(0, n) \qquad \begin{cases} x_1 \mapsto x_1 \\ x_2 \mapsto x_1 x_2 \end{cases} \qquad (3.15)$$

$$\mathcal{L}(0, -2) \hookrightarrow \mathcal{L}(0, -1) \qquad \begin{cases} x_1 \mapsto x_1 \\ x_2 \mapsto x_2^2 \end{cases} . \qquad (3.16)$$

Proof: For the first inflation, a simple generalization of Figure 2.20 is left to the reader. Figure 3.5 illustrates the isotopy for the second inflation. In both cases, one needs to use the belt trick when "pulling out" the subtemplate.   □

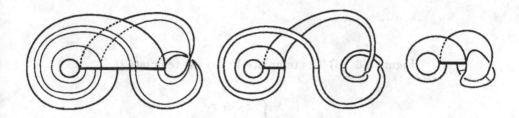

Figure 3.5: The template $\mathcal{L}(0, -2)$ is a subtemplate of $\mathcal{L}(0, -1)$.

The chain of inclusions among Lorenz-like templates implied by Proposition 3.2.1 is

$$\cdots \subset \mathcal{L}(0,4) \subset \mathcal{L}(0,2) \subset \mathcal{L}(0,0) \subset \mathcal{L}(0,-2) \subset \mathcal{L}(0,-4) \subset \cdots$$
$$\cap$$
$$\cdots \subset \mathcal{L}(0,5) \subset \mathcal{L}(0,3) \subset \mathcal{L}(0,1) \subset \mathcal{L}(0,-1) \subset \mathcal{L}(0,-3) \subset \cdots \qquad (3.17)$$

The templates $\mathcal{U}$ and $\mathcal{V}$

As a more intricate example of subtemplate structures, we turn to two deceptively simple templates first studied in [169] and later in [69].

Let $\mathcal{V}$ denote the embedded template of Figure 3.6(a), also introduced in Example 2.4.10. Let $\mathcal{U}$ denote the embedded template of Figure 3.6(b). Each template has two branch lines, ℓ_1 and ℓ_2, and four strips, labeled $x_1, \ldots, x_4$.

These templates are related in a fascinating way:

Proposition 3.2.2 *The following are isotopic template inflations:*

$$\mathfrak{F} : \mathcal{U} \hookrightarrow \mathcal{V} \qquad \begin{cases} x_1 \mapsto x_1 \\ x_2 \mapsto x_1 x_2 x_3 \\ x_3 \mapsto x_4 x_2 \\ x_4 \mapsto x_4 \end{cases} , \qquad (3.18)$$

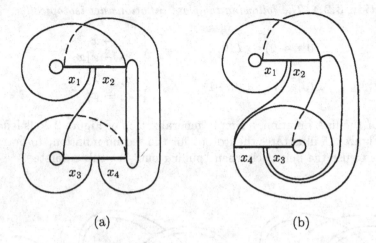

(a)   (b)

Figure 3.6: (a) The template $\mathcal{V}$; (b) the template $\mathcal{U}$.

$$\mathfrak{G} : \mathcal{V} \hookrightarrow \mathcal{U} \qquad \left\{ \begin{array}{l} x_1 \mapsto x_1 \\ x_2 \mapsto x_1 \\ x_3 \mapsto x_2 x_4 \\ x_4 \mapsto x_2 x_3 x_4 \end{array} \right. \qquad (3.19)$$

Proof: See the isotopies in Figures 3.7 and 3.8.   □

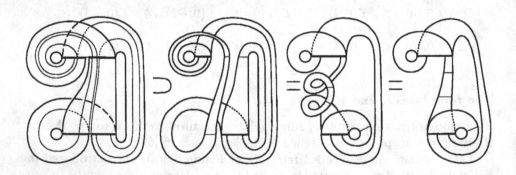

Figure 3.7: The template inflation $\mathfrak{F}$ acts isotopically.

Proposition 3.2.2 presents a puzzling situation: $\mathcal{U} \subset \mathcal{V}$ and $\mathcal{V} \subset \mathcal{U}$, and the inclusions occur in many different ways. By incorporating the symbolic approach to subtemplates of §2.4, we can track these various inclusions. For example, $\mathcal{U}$ and $\mathcal{V}$ display a symmetry which may be exploited to generalize the template inflations $\mathfrak{F}$ and $\mathfrak{G}$:

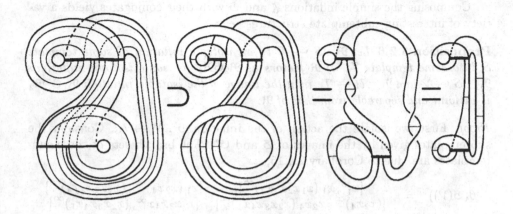

Figure 3.8: The template inflation $\mathfrak{G}$ acts isotopically.

Lemma 3.2.3 *The template inflation*

$$\chi : \begin{array}{l} \mathcal{U} \to \mathcal{U} \\ \mathcal{V} \to \mathcal{V} \end{array} \left\{ \begin{array}{l} x_1 \mapsto x_3 \\ x_2 \mapsto x_4 \\ x_3 \mapsto x_1 \\ x_4 \mapsto x_2 \end{array} \right. \tag{3.20}$$

takes each orbit to its mirror image.

Proof: The action of χ is to exchange the branch lines. As the only crossings in the templates of Figure 3.6 are at the branch lines, and these are of opposite sign, the inflation χ reverses the crossings of each template. □

Lemma 3.2.4 *Given any isotopic template inflation $\mathfrak{R}$ having either $\mathcal{U}$ or $\mathcal{V}$ as domain and either $\mathcal{U}$ or $\mathcal{V}$ as range, the conjugate inflation, $\mathfrak{R}^* = \chi \mathfrak{R} \chi$, is also isotopic.*

Proof: While the symbolic actions of χ and $\mathfrak{R}$ do not commute, the topological actions do. To see this, note that taking the mirror image commutes with the Reidemeister moves of Figure 1.3. Hence, topologically, $\mathfrak{R}^*$ acts as $\chi^2 \mathfrak{R}$. But, by Lemma 3.2.3, χ^2 is the identity, and $\mathfrak{R}^*$ acts as $\mathfrak{R}$: isotopically. □

Example 3.2.5 Conjugate inflations allow us to increase our "vocabulary" of inflations on the templates $\mathcal{U}$ and $\mathcal{V}$; e.g.,

$$\mathfrak{F}^* : \mathcal{U} \hookrightarrow \mathcal{V} \left\{ \begin{array}{l} x_1 \mapsto x_2 x_4 \\ x_2 \mapsto x_2 \\ x_3 \mapsto x_3 \\ x_4 \mapsto x_3 x_4 x_1 \end{array} \right. \tag{3.21}$$

Composing the simple inflations $\mathfrak{F}$ and $\mathfrak{G}$ with their conjugates yields a variety of interesting subtemplate structures: *e.g.*,

Proposition 3.2.6 *Let* $\mathfrak{R} : \mathcal{S} \hookrightarrow \mathcal{T}$ *be an isotopic inflation of some template* $\mathcal{S}$ *into some template* $\mathcal{T}$. *If* $\mathfrak{R}$ *factors as* $\mathfrak{R}_2\mathfrak{G}\mathfrak{R}_1$ *for some isotopic inflations* $\mathfrak{R}_1 : \mathcal{S} \hookrightarrow \mathcal{V}$ *and* $\mathfrak{R}_2 : \mathcal{U} \hookrightarrow \mathcal{T}$, *then the image of the isotopic inflation* $\mathfrak{R}_1\mathfrak{G}^*\mathfrak{R}_2$ *is disjoint and separable from that of* $\mathfrak{R}$.

Proof: First, we isolate the action of the inflation $\mathfrak{G} : \mathcal{V} \hookrightarrow \mathcal{U}$. Consider the subtemplates given by the images of $\mathfrak{G}$ and $\mathfrak{G}^*$. The branch sets of these subtemplates are, due to Corollary 2.4.15,

$$\beta(\mathfrak{G}(\mathcal{V})) \quad = \quad \begin{matrix} [x_1^\infty, x_1\,(x_1x_2x_3x_4)^\infty] & [x_1\,(x_2x_4)^\infty, (x_1x_2x_3x_4)^\infty] \\ [(x_2x_4)^\infty, x_2x_4\,(x_2x_3x_4x_1)^\infty] & [x_2x_3x_4x_1^\infty, (x_2x_3x_4x_1)^\infty] \end{matrix}$$

$$\beta(\mathfrak{G}^*(\mathcal{V})) \quad = \quad \begin{matrix} [(x_4x_2)^\infty, x_4x_2\,(x_4x_1x_2x_3)^\infty] & [x_4x_1x_2x_3^\infty, (x_4x_1x_2x_3)^\infty] \\ [x_3^\infty, x_3\,(x_3x_4x_1x_2)^\infty] & [x_3\,(x_4x_2)^\infty, (x_3x_4x_1x_2)^\infty] \end{matrix} \; .$$

We claim that the images of these two inflations are disjoint subtemplates of $\mathcal{U}$, except for their common boundary orbit $(x_1x_2x_3x_4)^\infty$. This may be shown by checking that certain shifts of $\beta(\mathfrak{G})$ (considered as "intervals" under $\lhd$) do not intersect shifts of $\beta(\mathfrak{G}^*)$ except at their common boundary and at branch lines. Though this is perhaps computationally tedious, it is a finite process which works when pictures fail.

However, the simplest proof is to carefully check that Figure 3.9(a) accurately represents the subtemplates in question, and that these are disjoint. In Figure 3.9(b), we crush out the transverse direction of the semiflow in each subtemplate, yielding a link of two graphs. From this, it is clear that these graphs, and hence the subtemplates, are separable.

It follows, then, that the images of $\mathfrak{R}_1\mathfrak{G}\mathfrak{R}_2$ and $\mathfrak{R}_1\mathfrak{G}^*\mathfrak{R}_2$ must also be disjoint and separable copies of $\mathcal{S}$ in $\mathcal{T}$. $\square$

Corollary 3.2.7 *Each template* $\mathcal{U}$ *and* $\mathcal{V}$ *contains a countable infinity of subtemplates isotopic to* $\mathcal{U}$ *and* $\mathcal{V}$ *which are completely disjoint and separable.*

Proof: Define the inflation $\mathfrak{A}_n$ to be $(\mathfrak{F}\mathfrak{G})\,(\mathfrak{F}\mathfrak{G}^*)^n$, for $n = 0, 1, \ldots$. The image of each $\mathfrak{A}_n$ is a subtemplate of $\mathcal{V}$ isotopic to $\mathcal{V}$ thanks to Proposition 3.2.2. We claim that the image of $\mathfrak{A}_n$ is disjoint and separable from the image of each $\mathfrak{A}_{n+k}$ for $k > 0$. To prove this, note that $\mathfrak{A}_{n+k}$ factors as

$$\mathfrak{A}_{n+k} = \left\{ \mathfrak{F}\mathfrak{G}\,(\mathfrak{F}\mathfrak{G}^*)^{k-1} \right\} (\mathfrak{F}\mathfrak{G}^*)^{n+1}, \tag{3.22}$$

so that the image of $\mathfrak{A}_{n+k}$ is contained in the image of $(\mathfrak{F}\mathfrak{G}^*)^{n+1}$. By Proposition 3.2.6, the images of $\mathfrak{A}_n$ and $(\mathfrak{F}\mathfrak{G}^*)^{n+1}$ are disjoint and separable, since they differ by changing one $\mathfrak{G}$ to $\mathfrak{G}^*$. Therefore each template, $\mathcal{V}$ and $\mathcal{U}$, contains infinitely many separable copies of itself (and of the other template). $\square$

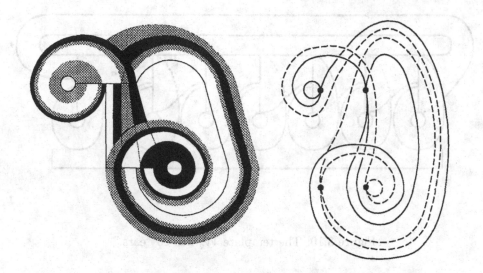

Figure 3.9: The subtemplates $\mathfrak{G}(\mathcal{V})$ and $\mathfrak{G}^*(\mathcal{V})$ (left) are disjoint and separable, as seen by reducing the subtemplates to embedded graphs (right).

3.2.2 A template containing all links

The embedded templates $\mathcal{U}$ and $\mathcal{V}$ of Corollary 3.2.7 entwine within one other in surprisingly complicated ways. We will exploit these subtemplate webs to answer basic questions about subtemplate structures. We begin with a solution to the existence problem for templates which are "universal" in the class of links.

Theorem 3.2.8 (Ghrist [69]) *The embedded template $\mathcal{V}$ contains representatives of every finite link as periodic orbits.*

The proof of Theorem 3.2.8 is the focal point of this chapter, and will be performed in steps.

We begin by examining a new family of templates, $\{\mathcal{W}_q; q \in \mathbb{Z}^+\}$, illustrated in Figure 3.10. Each $\mathcal{W}_q$ is an embedded q-fold cover of $\mathcal{V}$; that is, there are $2q$ "ears", or copies of the x_1 and x_3 strips. It is important to note that these ears alternate in crossing type — we denote them *positive*- and *negative*-type ears accordingly.

It is clear that there is a natural sequence of subtemplate inclusions $\mathcal{V} = \mathcal{W}_1 \subset \mathcal{W}_2 \subset \mathcal{W}_3 \subset \dots$. This increasing sequence is "large enough" to eventually contain any given link:

Proposition 3.2.9 *Given L an arbitrary link in S^3, an isotopic copy of L appears as a set of periodic orbits on the template $\mathcal{W}_q$ for q sufficiently large.*

Proof: Recall the braid group on N strands, B_N, from §1.1, generated by the elements $\sigma_i, i = 1 \dots N-1$. We construct "local" representatives of each generator

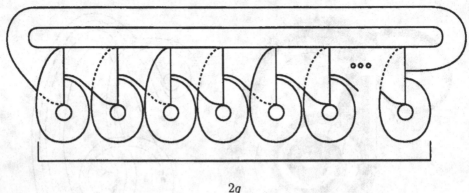

$$2q$$

Figure 3.10: The template $\mathcal{W}_q$ has $2q$ "ears."

(plus inverses) which live on $\mathcal{W}_q$ on a finite sequence of alternating ears. The arrangement of ears on $\mathcal{W}_q$ mimics the concatenation operation for the braid group.

In Figure 3.11, we show how to place the braid word $\sigma_1\sigma_2\ldots\sigma_k$ for any k on an ear with a positive crossing: the leftmost strand travels around the ear and is reinserted at an appropriate point. Similarly, we may place the word $\sigma_1^{-1}\sigma_2^{-1}\ldots\sigma_k^{-1}$ on an ear with a negative crossing. Assuming that some finite sequence of ears concatenated together yields the generators σ_j and σ_j^{-1} for all $j < k$, form the generator σ_k via concatenation:

$$\sigma_k = (\sigma_{k-1}^{-1})\ldots(\sigma_2^{-1})(\sigma_1^{-1})(\sigma_1\sigma_2\ldots\sigma_{k-1}\sigma_k). \qquad (3.23)$$

Hence, by induction, every σ_k and σ_k^{-1} fit on a finite sequence of alternating ears.

For $b \in B_N$ a braid on N strands, we may place the closed braid $\bar{b}$ on $\mathcal{W}_q$ for some (perhaps very large) q by piecing together the N-strand generators above on a finite sequence of alternating ears, then "connecting" the top and bottom. More specifically, since each component of the link can be given a sequence in some Markov structure for $\mathcal{W}_q$ (though this would be messy to do in practice), that orbit must exist on the template. We must be careful, however, that no two components of the closed braid have the same symbol sequence; else, they will not be distinct orbits on $\mathcal{W}_q$. To avoid this, note that since only one strand of the braid goes around an ear in the generators we use, it is sufficient to ensure that every strand of $\bar{b}$ goes around at least one ear. This may be done by appending the word $\sigma_{N-1}\sigma_{N-1}^{-1}$ to b: this does not change the braid element and hence the isotopy class of the resulting N-braid on $\mathcal{W}_q$. $\qquad\square$

Since $\mathcal{W}_q \subset \mathcal{W}_{q+1} \subset \ldots$ eventually contains any given link, our strategy is to show that *reverse* subtemplate inclusions also hold: $\mathcal{W}_q \subset \mathcal{W}_{q-1} \subset \ldots \subset \mathcal{V}$.

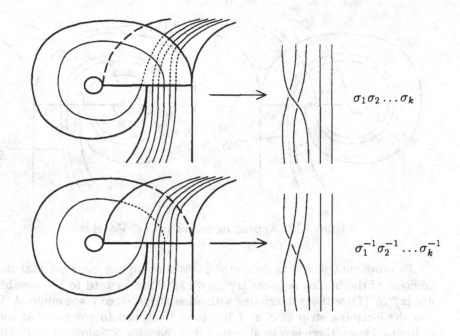

$\sigma_1\sigma_2\ldots\sigma_k$

$\sigma_1^{-1}\sigma_2^{-1}\ldots\sigma_k^{-1}$

Figure 3.11: The braid words $\sigma_1\sigma_2\ldots\sigma_k$ and $\sigma_1^{-1}\sigma_2^{-1}\ldots\sigma_k^{-1}$ fit on the ears of $\mathcal{W}_q$.

To find a copy of $\mathcal{W}_q$ within $\mathcal{V}$, we develop a type of surgery for subtemplates of $\mathcal{V}$. We denote the following procedure *appending an ear*.

Lemma 3.2.10 *Let $\mathcal{S} \subset \mathcal{V}$ be a subtemplate of $\mathcal{V}$ and let $I = [\partial^\ell(I), \partial^r(I)]$ be the component of $\mathcal{S} \cap \ell_1(\mathcal{V})$ which is minimal among all such intersections with respect to the $\lhd$ ordering on the upper branch line. If $\partial^\ell(I) \neq x_1^\infty$, then $\mathcal{S}$ is contained in a subtemplate $\mathcal{S}^+ \subset \mathcal{V}$ and this template $\mathcal{S}^+$ is isotopic to $\mathcal{S}$ except for the addition of an unknotted ear along I. Moreover, the subtemplate $\mathcal{S}^+$ contains the orbit $\partial_4^\ell(\mathcal{V})$.*

Proof: The subtemplate $\mathcal{S}$ is completely determined by its branch set $\beta(\mathcal{S})$, see Definition 2.4.2. That is, given $\beta(\mathcal{S})$, the subtemplate $\mathcal{S}$ is uniquely defined by flowing each branch segment forwards until it completely covers a collection of two or more branch segments. We specify the new subtemplate $\mathcal{S}^+$ by modifying $\beta(\mathcal{S})$.

Construct $\beta(\mathcal{S}^+)$ as follows: begin with $\beta(\mathcal{S}) \cup [x_1^\infty, x_1\partial^r(I)] \cup I$. This has the effect of adding a new strip which goes once around the x_1 strip and attaches at the new branch line $[x_1^\infty, \partial^r(I)]$. Then, to form a well-defined subtemplate, whenever an endpoint of some interval of $\beta(\mathcal{S}^+)$ ends in $\partial^\ell(I)$, replace this string with the string x_1^∞. This has the effect of "thickening" the portion of $\mathcal{S}^+$ which comes in along the x_4 strip of $\mathcal{V}$: see Figure 3.12.

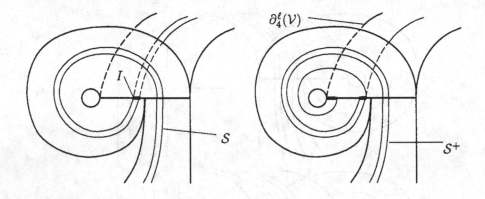

Figure 3.12: Appending an ear to $S \subset V$ yields S^+.

To prove that $\beta(S^+)$ as defined yields a subtemplate, we note that the the addition of the branch segment $[x_1^\infty, x_1 \partial^r(I)]$ flows forward to the new branch line $[x_1^\infty, \partial^r(I)]$ without interfering with other strips, since I was minimal. What was the incoming strip of S at I has been thickened to cover x_1^∞ at the left endpoint; hence, there is a local branch line chart for S^+ along $[x_1^\infty, \partial^r(I)]$.

Finally, we note that the appended ear is unknotted and "separable" from the rest of the subtemplate since the core orbit x_1^∞ is a separable unknot. Also, in thickening up the incoming strip along x_4, we include the orbit $\partial_4^\ell(V)$ in S^+ (this fact will be used later in Theorem 3.2.14).   □

The appended ear along I is a positive ear, since the crossing of the ear over the rest of the subtemplate is in the positive sense; similarly, negative ears may be added at the lower branch line:

Lemma 3.2.11 *Let $S \subset V$ be a subtemplate of V and let $I = [\partial^\ell(I), \partial^r(I)]$ be the component of $S \cap \ell_2(V)$ which is minimal among all such intersections with respect to the $\lhd$ ordering. If $\partial^\ell(I) \neq x_1^\infty$, then S is contained in a subtemplate $S^- \subset V$ and this template S^- is isotopic to S except for the addition of an unknotted ear along I. Moreover, the subtemplate S^- contains the orbit $\partial_2^\ell(V)$.*

Proof: Apply the symmetry map χ to V, taking the subtemplate S to its mirror image S^* as per Lemma 3.2.3. The segment $\chi(I) \subset \ell_1$ then satisfies the hypotheses of Lemma 3.2.10, and one may append an ear to $\chi(S)$ to obtain a subtemplate $(S^*)^+$ having an appended positive ear. Again applying χ to V takes this subtemplate to its mirror image: a subtemplate isotopic to S with a negative (the mirror image of a positive) ear appended along $\chi^2(I) = I \subset \ell_2$. This template contains the orbit $\partial_2^\ell(V) = \chi\left[\partial_4^\ell(V)\right]$ as an orbit.   □

To build copies of W_q as subtemplates of V, we must find a way to map V inside of itself isotopically so as to avoid the x_1^∞ and x_3^∞ boundaries (*e.g.*, the isotopic renormalization $\mathfrak{D}$ of Example 2.4.16 will not do). Then, we may append positive and negative ears in such a way that the resulting template is, say, isotopic to $\mathcal{W}_2$, and an iterative procedure may be used to build successively larger subtemplates isotopic to $\mathcal{W}_q$. We begin with the appropriate renormalization on $\mathcal{V}$ which keeps track of certain orbits for the iterative procedure later:

Proposition 3.2.12 *The inflation* $\mathfrak{H} \equiv \mathfrak{F}^* \mathfrak{G} \mathfrak{F} \mathfrak{G}^*$ *takes* $\mathcal{V} \hookrightarrow \mathcal{V}$ *isotopically. Among all points of* $\mathfrak{H}(\mathcal{V}) \cap \ell_1(\mathcal{V})$, *the* $\lhd$-*minimal point is contained in the orbit* $\mathfrak{H}(\partial_2^\ell(\mathcal{V}))$.

Proof: The symbolic action of $\mathfrak{H}$ is

$$\mathfrak{H} \equiv \mathfrak{F}^* \mathfrak{G} \mathfrak{F} \mathfrak{G}^* : \mathcal{V} \hookrightarrow \mathcal{V} \quad \begin{cases} x_1 \mapsto x_2 x_3^2 x_4 x_1 (x_2 x_4)^2 x_2 x_3 x_4 x_1 \\ x_2 \mapsto x_2 x_3^2 x_4 x_1 (x_2 x_4)^3 x_2 x_3 x_4 x_1 \\ x_3 \mapsto x_2 x_3^2 x_4 x_1 x_2 x_4 \\ x_4 \mapsto x_2 x_3^2 x_4 x_1 x_2 x_4 \end{cases} . \quad (3.24)$$

That this inflation is isotopic follows from Proposition 3.2.2. To show which point in the image of $\mathcal{V}$ is $\lhd$-minimal in the upper branch line ℓ_1, it is sufficient to check the image of the boundary of $\mathcal{V}$. This boundary, $\partial(\mathcal{V})$, is given implicitly in Equation (2.23) — we first recall this information:

$$\partial(\mathcal{V}) = \begin{cases} \partial_1^\ell(\mathcal{V}) = x_1^\infty \\ \partial_1^r(\mathcal{V}) = x_1 (x_2 x_4)^\infty \\ \partial_2^\ell(\mathcal{V}) = x_2 x_3^\infty \\ \partial_2^r(\mathcal{V}) = (x_2 x_4)^\infty \\ \partial_3^\ell(\mathcal{V}) = x_3^\infty \\ \partial_4^r(\mathcal{V}) = x_3 (x_4 x_2)^\infty \\ \partial_5^\ell(\mathcal{V}) = x_4 x_1^\infty \\ \partial_5^r(\mathcal{V}) = (x_4 x_2)^\infty \end{cases} . \quad (3.25)$$

Next, compute the image of the endpoints $\partial_i^{\ell/r}(\mathcal{V})$ under the inflation $\mathfrak{H}$:

$$\mathfrak{H} \ : \quad \mathcal{V} \hookrightarrow \mathcal{V} \quad\quad\quad (3.26)$$

$$\partial_1^\ell(\mathcal{V}) \ \mapsto \ \left(x_2 x_3^2 x_4 x_1 (x_2 x_4)^2 x_2 x_3 x_4 x_1 \right)^\infty$$

$$\partial_1^r(\mathcal{V}) \ \mapsto \ x_2 x_3^2 x_4 x_1 (x_2 x_4)^2 x_2 x_3 x_4 x_1 \left(x_2 x_3^2 x_4 x_1 (x_2 x_4)^3 x_2 x_3 x_4 x_1 x_2 x_3^2 x_4 x_1 x_2 x_4 \right)^\infty$$

$$\partial_2^\ell(\mathcal{V}) \ \mapsto \ x_2 x_3^2 x_4 x_1 (x_2 x_4)^3 x_2 x_3 x_4 x_1 \left(x_2 x_3^2 x_4 x_1 x_2 x_4 \right)^\infty$$

$$\partial_2^r(\mathcal{V}) \ \mapsto \ \left(x_2 x_3^2 x_4 x_1 (x_2 x_4)^3 x_2 x_3 x_4 x_1 x_2 x_3^2 x_4 x_1 x_2 x_4 \right)^\infty$$

$$\partial_3^\ell(\mathcal{V}) \ \mapsto \ \left(x_2 x_3^2 x_4 x_1 x_2 x_4 \right)^\infty$$

$$\partial_3^r(\mathcal{V}) \ \mapsto \ x_2 x_3^2 x_4 x_1 x_2 x_4 \left(x_2 x_3^2 x_4 x_1 x_2 x_4 x_2 x_3^2 x_4 x_1 (x_2 x_4)^3 x_2 x_3 x_4 x_1 \right)^\infty$$

$$\partial_4^\ell(\mathcal{V}) \ \mapsto \ x_2 x_3^2 x_4 x_1 x_2 x_4 \left(x_2 x_3^2 x_4 x_1 (x_2 x_4)^2 x_2 x_3 x_4 x_1 \right)^\infty$$

$$\partial_4^r(\mathcal{V}) \ \mapsto \ \left(x_2 x_3^2 x_4 x_1 x_2 x_4 x_2 x_3^2 x_4 x_1 (x_2 x_4)^3 x_2 x_3 x_4 x_1 \right)^\infty .$$

From (3.24), the image of the first x_2 in $\partial_2^\ell(\mathcal{V})$ contains two x_1 symbols. We claim that a shift of the image of $\partial_2^\ell(\mathcal{V})$ to one of these two x_1 symbols is $\lhd$-minimal in $\ell_1(\mathcal{V})$ among all shifts of the image of every other endpoint of $\beta(\mathcal{V})$

which begin with x_1. That this is so is a simple matter of choosing the shift of the image of $\partial_2^\ell(\mathcal{V})$ which is $\lhd$-minimal in $\beta_1(\mathcal{V})$ and then comparing this to all such shifts of the other endpoints $\mathfrak{H}(\partial_i^{\ell/r}(\mathcal{V}))$. Using the $\lhd$-ordering, this can be done by hand or (more conveniently) by computer. In this manner, we calculate that

$$\sigma^{14}\mathfrak{H}(\partial_2^\ell(\mathcal{V})) = x_1 \left(x_2 x_3^2 x_4 x_1 x_2 x_4\right)^\infty \tag{3.27}$$

is $\lhd$-minimal among all other orbits in the image of $\mathfrak{H}$ in $\ell_1(\mathcal{V})$, where σ denotes the shift operator. □

Note that the $\lhd$-minimal point in $\mathfrak{H}(\mathcal{V})$ on ℓ_1 is *not* x_1^∞ — thus, we may use this renormalization to append positive ears. The conjugate inflation will be used to append negative ears:

Proposition 3.2.13 *The inflation* $\mathfrak{H}^* \equiv \mathfrak{F}\mathfrak{G}^*\mathfrak{F}^*\mathfrak{G}$ *takes* $\mathcal{V} \hookrightarrow \mathcal{V}$ *isotopically. Among all points of* $\mathfrak{H}^*(\mathcal{V}) \cap \ell_2(\mathcal{V})$*, the* $\lhd$*-minimal point is contained in the orbit* $\mathfrak{H}^*(\partial_4^\ell(\mathcal{V}))$.

Proof: Since $\mathfrak{H}$ is isotopic, so is the conjugate $\mathfrak{H}^*$ via Lemma 3.2.4. Apply χ to Equation (3.27) to show that

$$\chi\sigma^{14}\mathfrak{H}(\partial_2^\ell(\mathcal{V})) = \chi \left\{ x_1 \left(x_2 x_3^2 x_4 x_1 x_2 x_4\right)^\infty \right\} \tag{3.28}$$

is $\chi(\lhd)$-minimal in $\chi(\ell_1(\mathcal{V}))$; after an application of Lemma 3.2.3 and the fact that χ commutes with the shift operator σ,

$$\sigma^{14}\chi\mathfrak{H}(\partial_2^\ell(\mathcal{V})) = x_3 \left(x_4 x_1^2 x_2 x_3 x_4 x_2\right)^\infty \tag{3.29}$$

is $\lhd$-minimal in $\ell_2(\mathcal{V})$. Now insert χ^2 in the domain. Since χ is involutive, we have shown that

$$\sigma^{14}\chi\mathfrak{H}\chi(\chi\partial_2^\ell(\mathcal{V})) = \sigma^{14}\mathfrak{H}^*(\partial_4^\ell(\mathcal{V})) = x_3 \left(x_4 x_1^2 x_2 x_3 x_4 x_2\right)^\infty, \tag{3.30}$$

is $\lhd$-minimal in $\ell_2(\mathcal{V})$. □

We may now complete the major step in the proof of Theorem 3.2.8.

Theorem 3.2.14 *The template* $\mathcal{W}_q$ *appears as a subtemplate of* $\mathcal{V}$ *for all* $q > 0$.

Proof: As we will be working with a series of distinct copies of the template $\mathcal{V}$, we introduce some notation. Let $\{\mathcal{V}^i\}$ denote a sequence of *distinct copies* of the embedded template $\mathcal{V}$ — each is embedded in a different copy of S^3. Construct an alternating sequence of templates and isotopic inflations:

$$\mathcal{V}^1 \xrightarrow{\mathfrak{H}} \mathcal{V}^2 \xrightarrow{\mathfrak{H}^*} \mathcal{V}^3 \xrightarrow{\mathfrak{H}} \mathcal{V}^4 \xrightarrow{\mathfrak{H}^*} \mathcal{V}^5 \xrightarrow{\mathfrak{H}} \mathcal{V}^6 \xrightarrow{\mathfrak{H}^*} \cdots \tag{3.31}$$

By Proposition 3.2.12, we may append a positive ear to $\mathfrak{H}(\mathcal{V}^1)$ in $\mathcal{V}^2$ along the image of $\partial_2^\ell(\mathcal{V}_1)$, creating the template denoted $\mathcal{W}_1^+ \subset \mathcal{V}^2$. This subtemplate contains the orbit $\partial_4^\ell(\mathcal{V}^2)$. By mapping $\mathcal{V}^2$ into $\mathcal{V}^3$ via $\mathfrak{H}^*$, we push $\mathcal{W}_1^+$ to a deeper isotopic copy within $\mathcal{V}^3$. A negative ear may then be appended to $\mathfrak{H}^*(\mathcal{W}_1^+) \subset \mathcal{V}^3$ along $\mathfrak{H}^*(\partial_4^\ell(\mathcal{V}^2))$ according to Proposition 3.2.13. Since the negative ear is appended along an interval having endpoint on $\mathfrak{H}^*(\partial_4^\ell(\mathcal{V}^2))$, the appended negative ear "precedes" the formerly appended positive ear (in the sense of the flow-direction), yielding a subtemplate of $\mathcal{V}^3$ isotopic to $\mathcal{W}_2$: see Figure 3.13.

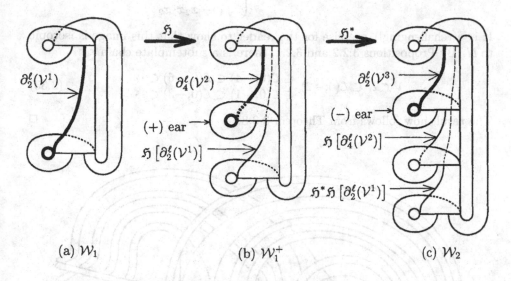

(a) $\mathcal{W}_1$ (b) $\mathcal{W}_1^+$ (c) $\mathcal{W}_2$

Figure 3.13: The steps in building $\mathcal{W}_q$.

We now have the template $\mathcal{V}^3$ containing a subtemplate isotopic to $\mathcal{W}_2$ which contains the orbit $\partial_2^\ell(\mathcal{V}^3)$. Since $\mathcal{V}^3$ is again an isotopic copy of $\mathcal{V}^1$ with $\partial_2^\ell(\mathcal{V}^3)$ corresponding to $\partial_2^\ell(\mathcal{V}^1)$, we may now iterate the procedure. Map $\mathcal{V}^3$ into $\mathcal{V}^4$ via $\mathfrak{H}$, append a positive ear to the image of $\mathcal{W}_2$ to obtain $\mathcal{W}_2^+$, then apply $\mathfrak{H}^*$ and append a negative ear to the image of $\mathcal{W}_2^+$ to produce $\mathcal{W}_3$. Since all the inflations involved are isotopic, we continue to carry the completed $\mathcal{W}_i$ along isotopically as we append additional ears. Thus, we can embed $\mathcal{W}_q$ in $\mathcal{V}$ for arbitrary q. $\qquad\qquad\qquad\qquad\qquad\qquad\qquad\qquad\qquad\qquad\qquad\qquad\qquad$ $\square$

Proof of Theorem 3.2.8: According to Theorem 1.1.13, any link may be represented as some closed braid. By Proposition 3.2.9, this closed braid must appear on $\mathcal{W}_q$ for q sufficiently large; hence, by Theorem 3.2.14, this link lives on $\mathcal{V}$. $\square$

3.2.3 Universal templates

Definition 3.2.15 A *universal template* is a template $\mathcal{T} \subset S^3$ among whose periodic orbits are representatives of every link type.

From Theorem 3.2.8, we may show the abundance of universal templates.

Proposition 3.2.16 *The Lorenz-like templates $\mathcal{L}(0,n)$ are universal for $n < 0$*

Proof: In Figure 3.14, we show the image of the inflation

$$\mathfrak{L}: \mathcal{U} \hookrightarrow \mathcal{L}(0,-2) \qquad \begin{cases} x_1 \mapsto x_1 \\ x_2 \mapsto x_1^3 x_2 \\ x_3 \mapsto x_2 x_1 \\ x_4 \mapsto x_2 x_1 x_2 \end{cases} . \qquad (3.32)$$

It is a (challenging!) exercise for the reader to show that this image is isotopic to $\mathcal{U}$. By Propositions 3.2.2 and 3.2.1, there is a subtemplate chain

$$\mathcal{V} \subset \mathcal{U} \subset \mathcal{L}(0,-2) \begin{array}{l} \subset \mathcal{L}(0,-4) \subset \mathcal{L}(0,-6) \subset \cdots \\ \subset \mathcal{L}(0,-1) \subset \mathcal{L}(0,-3) \subset \cdots \end{array} . \qquad (3.33)$$

The result now follows from Theorem 3.2.8.   □

Figure 3.14: The template $\mathcal{U}$ is a subtemplate of $\mathcal{L}(0,-2)$.

Given some embedded template, it is often relatively easy to recognize a Lorenz-like subtemplate; hence, we have a useful test for identifying universal templates.

Corollary 3.2.17 *Sufficient conditions for a template $\mathcal{T} \subset S^3$ to be universal are*

1. There is a two-component unlink on $\mathcal{T}$; that is, there exist two separable unknots.

2. One component of the unlink is untwisted; the other is twisted with $\tau \neq 0$ twists.

3. The two unknots intersect some branch line of $\mathcal{T}$ in two adjacent points (recall Definition 3.1.5) The sign of the branch line crossing between these two points must be opposite that of the twist τ.

Proof: Let $\mathbf{a}^\infty$ and $\mathbf{b}^\infty$ denote the adjacent points in Σ_{ℓ_j}, with $\mathbf{b}^\infty$ denoting the orbit with twist τ. For $\tau < 0$, the template inflation

$$\mathcal{L}(0,\tau) \hookrightarrow \mathcal{T} \qquad \left\{ \begin{array}{l} x_1 \mapsto \mathbf{a} \\ x_2 \mapsto \mathbf{b} \end{array} \right. \qquad (3.34)$$

is isotopic, since $\mathbf{a}^\infty$ and $\mathbf{b}^\infty$ are an unlink and there is agreement between twisting and branch line orientation. For $\tau > 0$, the same symbolic map sends the mirror image of $\mathcal{L}(0, -\tau)$ into $\mathcal{T}$ isotopically. However, the mirror image of a universal template is also universal.   □

We may use Corollary 3.2.17 to show that certain hyperbolic flows on S^3 contain all links as periodic orbits: e.g.,

Proposition 3.2.18 The suspension of the Plykin map, given in Example 2.1.6, when embedded in S^3 in the "standard" way, yields a flow having all link-types as periodic orbits.

Proof: Recall the Plykin attractor Λ_P described in Example 2.1.6. The inverse limit construction of Williams implies that we can collapse the attractor for the map to a branched one-manifold which suspends to a semiflow on a branched two-manifold[3]. In Figure 3.15, we show two periodic orbits in the suspension of the Plykin graph. The first orbit, γ_a, has period one, is untwisted, and is clearly separable from all other orbits. The second orbit, γ_b, is an unknot.

It is not hard to see that γ_b must be a twisted orbit; however, even if it were not, we could use Proposition 3.1.19 and Corollary 3.1.10 to show the existence of another orbit which is a twisted unknot separable from γ_a. Finally, we do not need to know the sign of the twist, since on the "branch line" (the graph Γ_P), the orbit γ_a is adjacent to a point of γ_b on either side, so it has branch line crossings of both types; hence, by Corollary 3.2.17, the periodic orbits of this flow contain all link types.   □

Corollary 3.2.17 is genuinely useful in this instance, since it is very difficult to draw an accurate picture of the entire template for the suspended Plykin attractor. The Plykin attractor is the simplest hyperbolic planar attractor. We have examined a few other examples and have managed to show that these also give rise to universal templates: we do not know of an example which does not.

---

[3]Though the suspension of the Plykin graph does not satisfy the definition of a template, it may be thought of as a template with the boundaries sewn together.

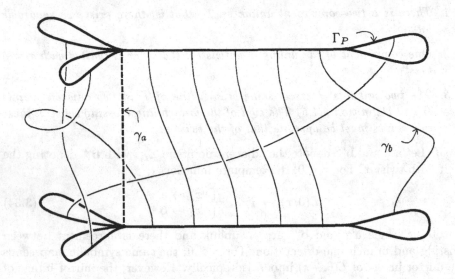

Figure 3.15: The suspension of the Plykin attractor — top are bottom are identified. Two orbits, γ_a and γ_b, form a "spine" for a universal subtemplate.

Corollary 3.2.19 *There exists a structurally stable vector field on S^3 such that the induced flow on S^3 contains closed orbit representatives of all knot and link types.*

Proof: The Plykin map suspends to a flow on $D^2 \times S^1$ which is inwardly transverse on the boundary and has chain recurrent set consisting of three attracting periodic orbits and the suspended Plykin attractor. Complete the flow on S^3 by taking another $D^2 \times S^1$ having a single repelling periodic orbit as $\{0\} \times S^1$ and outwardly transverse at the boundary and gluing these two solid tori together to get S^3 (a more detailed treatment of this construction appears in §A.1). The resulting flow has a hyperbolic chain recurrent set and hence, by Theorem 1.2.14, is structurally stable to C^1 perturbations.   □

Remark 3.2.20 There are numerous examples of flows on S^3 having all link types as periodic orbits. In §4.4, we will show that flows arising from certain "simple" ordinary differential equations can be modeled with a universal template. In [69], it was shown that certain fibred knots, namely the figure-eight knot and the Borromean rings, have complement fibred by a fibration whose induced flow contains all links as orbits (recall §2.3.4). As an exercise, the reader may wish to find two orbits on the template for the Whitehead link complement, Figure 2.17, which satisfy the conditions of Corollary 3.2.17, showing that this also is a universal template.

The Lorenz-like templates are the simplest class of templates: they have two unknotted unlinked strips with one branch line. A complete classification of these templates into universal and non-universal would be useful, *cf.* Corollary 3.2.17. At this time, we can offer only the following:

Proposition 3.2.21 *For* $mn \geq 0$, *the Lorenz-like template* $\mathcal{L}(m,n)$ *is universal if and only if* m *or* n *is* 0 *and the other index is negative.*

Proof: Proposition 3.2.16 covers the case where one number is zero and the other is negative: we will show that all other cases with $mn \geq 0$ are not universal. In the case where m and n are both nonnegative, the template $\mathcal{L}(m,n)$ contains only positive crossings and therefore carries no knots with mixed crossings (such as the figure-eight knot). Next, consider the case where m and n are both negative. Let K_a and K_b be two distinct knots on $\mathcal{L}(m,n)$ which form a link. If $\mathcal{L}(m,n)$ were universal, there would be an infinite number of distinct choices for K_a and K_b which would span all possible linking numbers. We compute the linking number as one half the algebraic sum of the total number of crossings, C, as per Equation (1.2). The crossing number, C, can be decomposed into the sum

$$C = C_m + C_n + C_o, \tag{3.35}$$

where C_m equals the contribution due to the m half-twists along the x_1 strip, C_n equals the contribution due to the n half-twists along the x_2 strip, and C_o equals the number of crossings due to the overlap of the x_1 strip over the x_2 strip at the branch line. Denote by a_{ij} (resp. b_{ij}) the number of $x_i x_j$ blocks in the periodic itinerary of K_a (resp. K_b). *Example:* if $K_a = \left(x_1 x_2^2 x_1 x_2 \right)^\infty$, then $a_{11} = 0, a_{12} = a_{21} = 2$, and $a_{22} = 1$. We note that in all cases,

$$a_{12} = a_{21}, \qquad b_{12} = b_{21}. \tag{3.36}$$

We can calculate the crossing numbers C_m and C_n:

$$C_m = m(a_{11} + a_{12})(b_{11} + b_{12}) \leq 0, \qquad C_n = n(a_{21} + a_{22})(b_{21} + b_{22}) \leq 0. \tag{3.37}$$

We will maximize the crossing numbers in order to obtain upper bounds; hence, we assume that there is a minimal amount of negative twisting in the strips, thereby setting $m = n = -1$ in Equation (3.37). To maximize the overcrossing number C_o, we again assume that all potential crossings can in fact occur. This situation is displayed schematically in Figure 3.16, where the different strands do not represent the knots themselves, rather those portions of the knots which correspond to the numbers a_{11}, etc. From Figure 3.16, the crossing number is bounded above by

$$C_o \leq a_{12}b_{21} + a_{21}b_{12} + a_{11}b_{21} + a_{21}b_{11} + a_{22}b_{12} + a_{12}b_{22}. \tag{3.38}$$

Combining this with Equations (3.35) and (3.36) yields

$$
\begin{aligned}
C &\leq 2a_{12}b_{12} + a_{11}b_{12} + a_{12}b_{11} + a_{22}b_{12} + a_{12}b_{22} - a_{11}b_{11} - a_{11}b_{12} \\
&\quad - a_{12}b_{11} - a_{12}b_{12} - a_{12}b_{12} - a_{12}b_{22} - a_{22}b_{12} - a_{22}b_{22} \\
&\leq -(a_{11}b_{11} + a_{22}b_{22}) \\
&\leq 0
\end{aligned}
\tag{3.39}
$$

$$a_{11} \quad b_{11} \quad a_{12}\ b_{12} \qquad\qquad b_{21}\ a_{21}\ b_{22}\ a_{22}$$

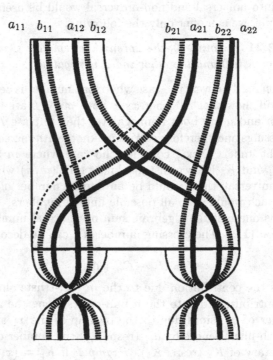

Figure 3.16: A schematic diagram of crossings on the template $\mathcal{L}(-1,-1)$: each strand labeled a_{ij} (resp. b_{ij}) represents a collection of a_{ij} (resp. b_{ij}) strands of the knot K_a (K_b) which begin on the strip x_i and end on the strip x_j.

Hence, the linking number $\ell k(K_a, K_b)$ is at most zero and $\mathcal{L}(m,n)$ cannot support all links.   □

We have classified the universal Lorenz-like templates in every case except $m < 0, n > 0$ (and *vice versa*). The linking number estimates in the proof do not yield the necessary results in the case when m and n are of mixed sign. We settle for the following:

Conjecture 3.2.22 *A Lorenz-like template $\mathcal{L}(m,n)$ supports all links if and only if either m or n is zero and the other index is negative.*

The most pressing problem concerning universal templates is to determine a simple set of necessary and sufficient conditions for universality. We conclude with two related conjectures.

Conjecture 3.2.23 *An embedded template $\mathcal{T} \subset S^3$ is universal if and only if it contains $\mathcal{V}$ as a subtemplate.*

Conjecture 3.2.24 An embedded template $\mathcal{T} \subset S^3$ is universal if and only if it contains a countable untwisted unlink: each component of which is an untwisted unknot, separable from all other components.

Conjecture 3.2.24 would give an obstruction to hyperbolic dynamics in flows. For example, the suspension of the identity map on D^2 has a countable untwisted unlink, yet, it does not support closed orbits of all knot types; hence it is not a hyperbolic system.

3.2.4 Where do all the knots live?

The topological richness of closed orbits on templates that we have examined in this section is at first mysterious. Given an innocuous looking template such as $\mathcal{V}$, it is hard to imagine what a very complicated knot (*e.g.*, the connected sum of a thousand trefoils) must look like on this template. As an addendum to this section, we give a quick computation illustrating how even a "simple" knot may require a rather complex presentation on a universal template.

The proof of Theorem 3.2.8 is constructive. So, in theory, we should be able to compute a representative of any given closed braid on $\mathcal{V}$. Consider the figure-eight knot, denoted K_8. This link in closed braid form has a presentation (in the standard generators) with three strands as $(\sigma_2 \sigma_1^{-1})^2$. To place this knot on $\mathcal{W}_q$ for some q, we write the generators σ_i in the form of Proposition 3.2.9:

$$(\sigma_2 \sigma_1^{-1})^2 = ()(\sigma_1^{-1})(\sigma_1 \sigma_2)(\sigma_1^{-1})()(\sigma_1^{-1})(\sigma_1 \sigma_2)(\sigma_1^{-1}), \qquad (3.40)$$

where the empty parentheses () denote positive ears that are not traversed in arranging K_8 on $\mathcal{W}_4$.

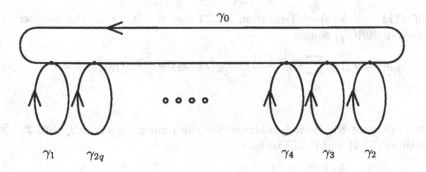

Figure 3.17: The spine of W_q, with fundamental loops labeled.

From the proof of Theorem 3.2.14, we know that $\mathcal{W}_4$, and hence K_8, live on $\mathcal{V}$. Although the proof does not supply a precise inflation from $\mathcal{W}_q$ to $\mathcal{V}$, the symbolic action of the construction is traceable in part. In Figure 3.17, we present the spine of the template $\mathcal{W}_q$, formed by crushing out the transverse direction to the semiflow. The generators of the fundamental group of $\mathcal{W}_q$ are labeled $\gamma_0, \gamma_1, \ldots, \gamma_{2q}$ in the order in which they are constructed within $\mathcal{V}$. By carefully following the proof of Theorem 3.2.14, one can track the images of these loops γ_i in $\mathcal{V}$ for the "simplest" copy of $\mathcal{W}_q$ in $\mathcal{V}$:

i	$\gamma_i \in \pi_1(\mathcal{W}_q)$
0	$(\mathfrak{H}^*\mathfrak{H})^{q-1}(x_2 x_4) = \left[\mathfrak{F}\mathfrak{G}^*(\mathfrak{F}^*\mathfrak{G})^2\mathfrak{F}\mathfrak{G}^*\right]^{q-1}(x_2 x_4)$
1	$(\mathfrak{H}^*\mathfrak{H})^{q-1}(x_1) = \left[\mathfrak{F}\mathfrak{G}^*(\mathfrak{F}^*\mathfrak{G})^2\mathfrak{F}\mathfrak{G}^*\right]^{q-1}(x_1)$
$i = 2k > 0$	$(\mathfrak{H}^*\mathfrak{H})^{q-k}(x_3) = \left[\mathfrak{F}\mathfrak{G}^*(\mathfrak{F}^*\mathfrak{G})^2\mathfrak{F}\mathfrak{G}^*\right]^{q-k}(x_3)$
$i = 2k+1 > 1$	$(\mathfrak{H}^*\mathfrak{H})^{q-k-1}\mathfrak{H}^*(x_1) = \left[\mathfrak{F}\mathfrak{G}^*(\mathfrak{F}^*\mathfrak{G})^2\mathfrak{F}\mathfrak{G}^*\right]^{q-k-1}\mathfrak{F}\mathfrak{G}^*\mathfrak{F}^*\mathfrak{G}(x_1)$

$$(3.41)$$

From this table, we could compute the symbol sequence of this representative of K_8 in $\mathcal{V}$; however, printing it out might take more room than our publisher wishes to spare. We merely compute the symbolic period, $i.e.$, the length of the repeating block of the periodic word.

The knot K_8 on $\mathcal{W}_4$ determines a word in $\pi_1(\mathcal{W}_4)$ in the γ_i generators — let n_i, $i = 0 \ldots 8$ denote the number of γ_i terms in this word. In other words, the link K_8 goes around the loop γ_i exactly n_i times. To compute $|\gamma_i|$, the symbolic length of the image of the loop γ_i in $\mathcal{V}$, we define a *symbolic growth matrix* for a template renormalization.

Definition 3.2.25 *Given a renormalization* $\mathfrak{R} : \mathcal{T} \hookrightarrow \mathcal{T}$, *where* $\mathcal{T}$ *has Markov partition* $\{x_1, x_2, \ldots, x_N\}$, *define the* growth matrix *of* $\mathfrak{R}$, $G_{\mathfrak{R}} \in M_N(\mathbb{Z}^+)$, *as follows:*

$$G_{\mathfrak{R}}(x_i, x_j) = \{\# \text{ of } x_i \text{ symbols in } \mathfrak{R}(x_j)\}\} \tag{3.42}$$

Lemma 3.2.26 *For any* $\mathfrak{R}$ *and* $\tilde{\mathfrak{R}} : \mathcal{T} \to \mathcal{T}$,

$$G_{\mathfrak{R}\tilde{\mathfrak{R}}} = G_{\mathfrak{R}}G_{\tilde{\mathfrak{R}}}. \tag{3.43}$$

Proof: This follows from Definition 3.2.25 and the fact that the number of x_i symbols in $\mathfrak{R}\tilde{\mathfrak{R}}(x_j)$ equals

$$G_{\mathfrak{R}\tilde{\mathfrak{R}}}(x_i, x_j) = \sum_k G_{\mathfrak{R}}(x_i, x_k)G_{\tilde{\mathfrak{R}}}(x_k, x_j) = \left[G_{\mathfrak{R}}G_{\tilde{\mathfrak{R}}}\right](x_i, x_j). \tag{3.44}$$

$\square$

We compute the growth matrices for the renormalizations $\mathfrak{H}$ and $\mathfrak{H}^*$ from Equations (3.24) and (3.20) to be

$$G_{\mathfrak{H}} = \begin{bmatrix} 2 & 2 & 1 & 1 \\ 4 & 5 & 2 & 2 \\ 3 & 3 & 2 & 2 \\ 4 & 5 & 2 & 2 \end{bmatrix} \qquad G_{\mathfrak{H}^*} = \begin{bmatrix} 2 & 2 & 3 & 3 \\ 2 & 2 & 4 & 5 \\ 1 & 1 & 2 & 2 \\ 2 & 2 & 4 & 5 \end{bmatrix}. \tag{3.45}$$

Hence, by using Lemma 3.2.26 and the information from (3.41), we can compute the growth matrix for the renormalization which takes each γ_i into $\mathcal{V}$. This information yields the length of the orbit γ_i in $\mathcal{V}$:

Example 3.2.27 To find $|\gamma_0|$ for $\mathcal{W}_4 \subset \mathcal{V}$, we look up γ_0 from (3.41) and note that it is the image of $(x_2 x_4)^\infty$ under $(\mathfrak{H}\mathfrak{H}^*)^3$. To count $|\gamma_0|$, the length, set $\vec{v} = [0, 1, 0, 1]^t$ and take the product $(G_{\mathfrak{H}} G_{\mathfrak{H}} \cdot)^3 \vec{v}$. Then sum all the entries of this column matrix — from Definition 3.2.25, this counts the number of x_2 and x_4 elements, giving the length $|\gamma_0|$.

i	n_i	$|\gamma_i|$	i	n_i	$|\gamma_i|$
0	3	3387648	5	0	839
1	0	1990365	6	1	77
2	1	1086485	7	1	7
3	1	99679	8	1	1
4	1	9145			

(3.46)

Finally, to obtain the length of the representative of the figure-eight knot K_8 in $\mathcal{V}$, a simple computation from (3.46) gives:

$$|K_8| = \sum_{i=0}^{8} n_i |\gamma_i| = 11,358,338, \qquad (3.47)$$

or, over eleven million. There are surely simpler representatives of K_8 on $\mathcal{V}$; however, the simplest may still be outside of the range in which one can draw it.[4]

This example illustrates that methods used in the proof of Theorem 3.2.8 extract relatively "deep" information from templates.

Remark 3.2.28 To compute upper bounds for the minimal length of a given knot type represented on $\mathcal{V}$, one need merely compute the Perron-Frobenius (*i.e.*, maximal) eigenvalue of the growth matrix $G_{\mathfrak{H}}$ — it is about 10.332. Then, given any knot K, write it in braid format which is compatible with the template $\mathcal{W}_q$, as done in Equation (3.40). Note that the length $|K|$ of the resulting braid word may be quickly estimated from any braid version of L via the procedure of Proposition 3.2.9. A (poor) lower bound for the length of an orbit representing K is then given by $(10.332)^{|K|-1}$, since then inflation $\mathfrak{H}$ (or $\mathfrak{H}^*$) must be applied $|K| - 1$ times to fit $\mathcal{W}_q$ with the braid form of K on it within $\mathcal{V}$. Applied to the figure eight example with braid length 8 (from Equation (3.40)), one gets an upper bound for the minimal length as 12,567,447 — off from our computed example by about ten percent.

3.3 Subtemplate structures

Although the results of Theorem 3.2.8 are exciting, we have, to some degree, drawn our conclusions too soon. The proof succeeds because it examines *subtemplate structures*, which carry the desired links, rather than examining the

---

[4]Two of us (MS, RG) tried *very* hard to find a copy of K_8 on $\mathcal{V}$ or $\mathcal{U}$ before Theorem 3.2.8 was discovered.

individual knots and links *per se*: the crucial step lies in showing $\mathcal{W}_q \subset \mathcal{V}$. We begin this section by resuming our study of the subtemplates of $\mathcal{V}$. The results of this line of inquiry will lead to generalizations of Theorem 3.2.8 and will suggest directions for further investigation along the lines of subtemplate structures. We do not present all the results in full detail: the interested reader should be able to fill in such as necessary.

We must distinguish between orientable and nonorientable cases, since an orientable template cannot contain any nonorientable subtemplates. In §3.3.1 and §3.3.2, we prove the existence of templates which are "universal" in the classes of orientable and nonorientable templates in that they contain isotopic copies of all [orientable] templates as subtemplates.

3.3.1   Orientable subtemplates

We begin with a generalization of the braid group structure of Definition 1.1.11 to a semigroup structure on braided templates. The generators of the semigroup are of three types:

1. $\sigma_i^{\pm}$, is a "flat ribbon" version of the generators for the braid group: the ith strip crosses over the $(i+1)$st in the positive sense. These elements are invertible;

2. $\tau_i^{\pm}$, is the trivial element (a collection of straight flat strips) with the ith strip given a half twist, either in the positive (τ_i) or negative (τ_i^{-1}) sense. These elements are invertible.

3. $\beta_i^{\pm}$, is a branch line chart with the ith and $(i+1)$st strips incoming, k outgoing strips[5], and either a positive (β_i) or a negative (β_i^{-1}) crossing at the branch line. *These generators are not inverses*, as branch lines cannot be cancelled under composition.

Figure 3.18 illustrates the generators.

The following result is obvious, and implicit in the proof of Theorem 3.1.2 [58]:

Lemma 3.3.1 *The set* $\{\sigma_i^{\pm}, \beta_i^{\pm}, \tau_i^{\pm}\}$ *generates the class of braided templates.*

With the braided template semigroup playing the role of B_N in Theorem 3.2.8, we may generalize this result to:

Theorem 3.3.2 *The template* $\mathcal{V}$ *contains every embedded orientable template* S *as a subtemplate. Furthermore, these may be chosen so as to be disjoint and separable.*

Proof: Recall Theorem 3.2.14 — $\mathcal{V}$ contains $\mathcal{W}_q$ for all q. The strategy of the proof of Theorem 3.2.8 was to show that any given closed braid can be fitted

---

[5]For simplicity, we suppress reference to the number of strips involved, which varies throughout the braid presentation, in our notation.

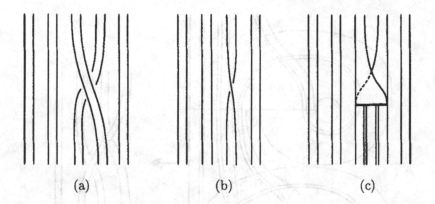

(a)   (b)   (c)

Figure 3.18: The generators for the braided template semigroup: (a) σ_i; (b) τ_i; (c) β_i.

onto some $\mathcal{W}_q$. We use the semigroup for braided templates in analogous fashion to show that all orientable templates also live on $\mathcal{W}_q$, and hence on $\mathcal{V}$.

Consider S a template in S^3, presented as a flat braided template as per Theorem 3.1.2. To show that such a given template lives as a subtemplate of $\mathcal{W}_q$ for some q, we will express each generator as a subtemplate of a portion of $\mathcal{W}_q$; that is, on a finite sequence of alternating ears.

In Figure 3.19, we exhibit a portion of a subtemplate on a pair of positive and negative ears which corresponds to the generator $\sigma_1 \sigma_2 \ldots \sigma_k$ for any desired k. Note that the belt trick is used in concert with two ears of opposite sign to cancel the full twist induced by going around an ear. One constructs the generator $\sigma_1^{-1} \sigma_2^{-1} \ldots \sigma_k^{-1}$ in analogous fashion. To show that some finite product of these yields σ_j and σ_j^{-1} for any j, we follow the same argument as in Proposition 3.2.9.

To show that β_i and β_i^{-1} appear likewise, we turn to Figure 3.20, which contains a local picture of the generator β_i. The first i strips travel around a negative ear and then a positive ear (or vice versa for β_i^{-1}) in order to cancel the twisting and allow for a positive (negative resp.) crossing at the branch line.

Since an orientable braided template may always be made flat, we do not need to fit powers of τ_i on $\mathcal{W}_q$; hence, the entire generating set for braided orientable templates appears locally on a finite set of alternating positive and negative ears. Piecing together local submanifolds on $\mathcal{W}_q$ is always possible as long as the number of strips matches — after including all the crossings, branch lines, etc., one simply connects the top to the bottom strips in the standard way. Hence, given any template presented in these standard generators, one may construct for some q (perhaps very large) a subtemplate of $\mathcal{W}_q$ which is isotopic to the intended template.

The result then follows from Theorem 3.2.14. □

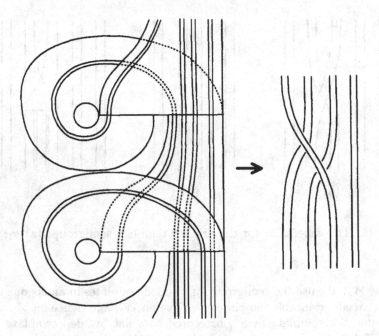

Figure 3.19: The braided-template word $\sigma_1 \sigma_2 \ldots \sigma_k$ lives on a pair of alternating ears.

Remark 3.3.3 Theorem 3.3.2 indicates that, among the class of orientable templates, $\mathcal{V}$ is not merely an example of an exceptional template: it (and all other such templates) truly deserves the title of *universal template*, since a template contains all orientable templates if and only if it contains $\mathcal{V}$.

Corollary 3.3.4 *The template $\mathcal{V}$ contains all evenly twisted links: that is, it contains all links indexed by the (even) twist of the local stable manifolds (see Definition 3.1.9).*

Proof: Given an indexed link L, where the components of L are indexed by the twist, build an orientable template $\mathcal{T}_L$ which contains the link L as its "spine." More specifically, form a connected graph from L by (arbitrarily) identifying points on components pairwise. Then, thicken the graph up to a template, adding branch lines at vertices and twisted strips along the edges as appropriate: *cf.* Figure 3.9(b). This template, which contains L as a set of periodic orbits, lives on $\mathcal{V}$ by Theorem 3.3.2.   □

Our next result shows that any orientable template may be embedded in S^3 as a universal template (and then some):

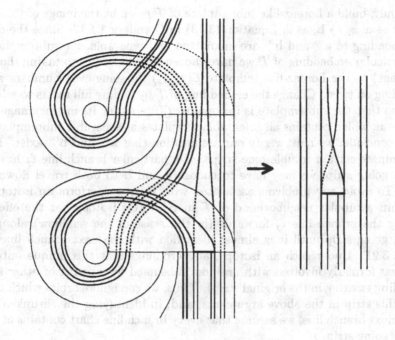

Figure 3.20: The generator β_i lives on a pair of alternating ears.

Theorem 3.3.5 *Any orientable template $\mathcal{T}$ may be embedded in S^3 so as to contain an isotopic copy of all orientable templates as disjoint separable subtemplates.*

Proof: Assume for the moment that, for some branch line ℓ_j, there exist two periodic points of Σ_{ℓ_j}, $\mathbf{a}^\infty$ and $\mathbf{b}^\infty$, such that each symbol x_i in the Markov partition of $\mathcal{T}$ appears at most once in the word $\mathbf{ab}$; thus, each strip of $\mathcal{T}$ contains at most one strand of the link $\{\mathbf{a}^\infty, \mathbf{b}^\infty\}$.

Re-embed $\mathcal{T}$ by changing the overcrossings of strips in the given planar presentation in the manner to be described: by the above condition, whenever the knots corresponding to $\mathbf{a}^\infty$ and $\mathbf{b}^\infty$ cross one another, they must do so on separate strips. Re-embed $\mathcal{T}$ so as to force the strip containing the orbit $\mathbf{a}^\infty$ to always be on top. In this embedding, then, the two knots are clearly separable.

Now restrict attention to those instances where the knot corresponding to $\mathbf{a}^\infty$ crosses itself: if $\mathbf{a}$ and $\mathbf{b}$ are chosen as above, this crossing must be due to a strip crossing over itself or another strip. Beginning at an arbitrary point on this orbit, follow along the direction of the flow — whenever there is a self-crossing, re-embed the strips so that the desigated point is on top. When finished, one has a knot which can be perturbed so as to have a unique local maximum: an unknot. Repeat this procedure for the knot $\mathbf{b}^\infty$, noting that one is not tampering with any previously re-embedded strips.

Finally, build a Lorenz-like subtemplate of $\mathcal{T}$ given by the image of the inflation $x_1 \mapsto \mathbf{a}, x_2 \mapsto \mathbf{b}$, as in Equation (3.13) in Corollary 3.1.17. Since the orbits
corresponding to $\mathbf{a}^\infty$ and $\mathbf{b}^\infty$ are unknotted and separable, the subtemplate in
the particular embedding of $\mathcal{T}$ we have chosen is isotopic (up to taking the mirror image) to the Lorenz-like template $\mathcal{L}(\tau_a, \tau_b)$ for some even numbers τ_a, τ_b,
depending on twist. Change the embedding of $\mathcal{T}$ by adding full twists to selected
strips so that the subtemplate is isotopic to $\mathcal{L}(0, -n)$ (or its mirror image), for
positive n, which contains all orientable templates as separable subtemplates.

To conclude, we must verify our assumption that $\mathbf{a}^\infty$ and $\mathbf{b}^\infty$ exist. First,
we eliminate certain troublesome strips. If a particular branch line ℓ_j has only
one outgoing strip, we may have to choose $\mathbf{a}$ and $\mathbf{b}$ to both travel down this
strip. To avoid any problems associated with this, we perform an isotopy on
$\mathcal{T}$ within a tubular neighborhood of $\mathcal{T}$ in S^3. This isotopy has the effect of
pushing the branch line ℓ_j forwards (in the sense of the semiflow) along the
one outgoing strip until it is almost identified with the next branch line: see
Figure 3.21. Under such an isotopy, any crossings that this unique outgoing
strip was formerly involved with are now subsumed by crossing of other strips
(including twisting in the original strip). Thus, we can ignore orbits which travel
down this strip in the above arguments, and, in identifying the shrunken strip
to the next branch line, we assume that every branch line chart contains at least
two outgoing strips.

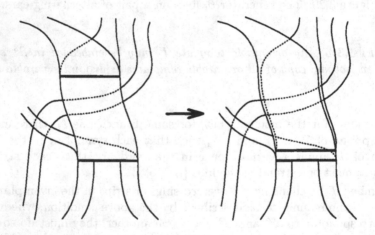

Figure 3.21: One can "eliminate" a single-outgoing strip by propagating the
branch line forwards.

Next, choose a finite admissible orbit $\mathbf{a}^\infty$. We claim that $\mathbf{a}$ may be chosen
such that the knot passes through each branch line at most once. Assume that
$\mathbf{a} = \mathbf{a}_1 \mathbf{a}_2 \mathbf{a}_3$, where $\mathbf{a}_2$ and $\mathbf{a}_3$ are words whose orbits begin from the same branch
line. Then, replace $\mathbf{a}$ with $\mathbf{a}_1 \mathbf{a}_3$: this is an admissible word since incoming strips
stretch over branch lines completely. Iterating this reduction on a word of finite
length is a terminal process.

Finally, we claim that **b** may be chosen similarly to have no symbols in common with **a**. Recall, we have modified $\mathcal{T}$ to have [in effect] at least two outgoing strips per branch line chart, and **a** intersects each branch line at most once. Beginning at some branch line of $\mathcal{T}$, choose an outgoing strip whose symbol is not part of **a** — this is always possible since there are more than two outgoing strips. This outgoing strip leads to another branch line. Repeat the process of choosing outgoing strips avoiding **a** until the branch line is repeated: this defines a periodic orbit **a'**. If **a'** and **a** have a branch line in common, this is the desired **b**. If not, repeat the process of choosing another periodic orbit **a''** — this algorithm may be repeated since there are again at least one incoming and outgoing strips per branch line on $\mathcal{T}$ minus the strips of **a** and **a'**. Since the Markov partition is finite, this is a finite process; hence, **a** and **b** may be chosen as above. $\square$

Corollary 3.3.6 *Any embedded orientable template $\mathcal{T}$ contains a (nonisotopically) embedded copy of every orientable template.*

3.3.2 Nonorientable subtemplates

The nonorientable case is quite a bit more subtle, but is solved in similar fashion. We leave the [numerous] details of the following theorem to the reader.

Theorem 3.3.7 *There exists a template which contains every embedded template S as a subtemplate.*

Idea of Proof: We begin with the template $\mathcal{L}(0,-2)$, which contains $\mathcal{V}$ via the inflation $\mathcal{LG}$, where $\mathcal{L}: \mathcal{U} \hookrightarrow \mathcal{L}(0,-2)$ is the inflation of Equation (3.32). Then, we append an extra ear to this template which is twisted and separable from the remainder of the template: see the template $\mathcal{Y}$ in Figure 3.22. Given any template, we then show that it may be obtained by first placing a similar orientable template on $\mathcal{L}(0,-2)$, then diverting some of the strips around the twisted appended ear of $\mathcal{Y}$ to produce the requisite nonorientable subtemplate.

Let S be an arbitrary embedded template in S^3. We briefly indicate how to place S in the appropriate form for being a subtemplate of $\mathcal{Y}$.

Step 1: Place S in braided form as per Theorem 3.1.2, and represent this template in the braid semigroup of Lemma 3.3.1.

Step 2: Factor this braid word so that there is a positive half-twist Δ on the first k strips, where Δ is the word

$$\Delta = \tau_1 \tau_2 \cdots \tau_k \left(\prod_{i=1}^{k-1} \sigma_1 \sigma_2 \cdots \sigma_i \right), \tag{3.48}$$

followed by a braid word having no $\tau_i^{\pm}$ terms.

Step 3: For S braided into the word above, let $\bar{S}$ denote the flat orientable template given by removing the initial word Δ from the braid word. Map $\bar{S}$

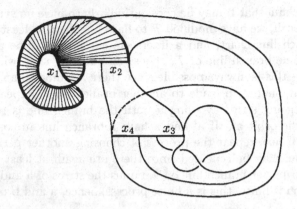

Figure 3.22: The template $\mathcal{Y}$ contains all templates as subtemplates.

into $\mathcal{L}(0, -2) \subset \mathcal{Y}$ isotopically via the inflation $\mathcal{L}\mathfrak{G}\mathfrak{I}$, where $\mathfrak{I} : \tilde{S} \hookrightarrow \mathcal{V}$ is the inflation from the proof of Theorem 3.2.8 and $\mathcal{L} : \mathcal{U} \hookrightarrow \mathcal{L}(0, -2)$ is the inflation from Equation (3.34).

Step 4: Now, by carefully tracking the placement of the first k strips in $S \subset \mathcal{Y}$, modify the inflation in the appropriate manner to "divert" the leftmost k strips of S on $\mathcal{Y}$ to instead make a loop around the appended twisted x_1-ear. This has the effect of inserting Δ into the braid word for $\tilde{S}$ at the beginning. This new template is the original S by Step 3.   □

Corollary 3.3.8 *The template $\mathcal{Y}$ contains isotopic copies of links with arbitrary twist type.*

Proof: See the proof of Corollary 3.3.4.   □

Remark 3.3.9 Note that although $\mathcal{Y}$ contains all embedded templates as disjoint subtemplates, these may not be chosen so as to be mutually unlinked in the present construction, since there is a linking induced by the trip about the twisted ear. We believe that this is unavoidable: *i.e.*, no embedded template contains disjoint unlinked copies of all embedded templates as subtemplates.

We do not believe that the results of Theorem 3.3.5 hold for nonorientable templates: that is, we do not believe it is possible to re-embed, say, the horseshoe template $\mathcal{L}(0, 1)$ in such a way that it contains copies of *every* embedded template, or even the orientable ones. A related, though weaker statement is however true:

Proposition 3.3.10 *Any embedded non-orientable template T contains a (non-isotopically) embedded copy of all templates.*

Proof: If $\mathcal{T}$ is nonorientable, we must construct an inflation from the template $\mathcal{Y}$ of Figure 3.22 into $\mathcal{T}$. Take $\mathbf{a}^\infty$ twisted and $\mathbf{b}^\infty$ untwisted with the pair adjacent. Then consider $\mathbf{c} = \mathbf{a}^2\mathbf{b}$: $\mathbf{c}^\infty$ is an untwisted orbit with $\mathbf{a}^\infty$ and $\mathbf{c}^\infty$ adjacent and $(\mathbf{ba}^2)^\infty$ and $\mathbf{b}$ adjacent. Hence, there is a well-defined template inflation,

$$\mathfrak{R}: \mathcal{Y} \hookrightarrow \mathcal{T} \qquad \begin{cases} x_1 \mapsto \mathbf{a} \\ x_2 \mapsto \mathbf{a}^2 \\ x_3 \mapsto \mathbf{b} \\ x_4 \mapsto \mathbf{b} \end{cases}. \tag{3.49}$$

As this inflation is nonisotopic, we have a different embedding of all the subtemplates of $\mathcal{Y}$ into $\mathcal{T}$. □

Ostensibly, it seems surprising that the template for the Whitehead link complement (Figure 2.17) embeds in the horseshoe template $\mathcal{L}(0, 1)$.

Remark 3.3.11 Although the results of this section are exciting, they may also be cause for concern in certain applications: recall from §2.3.5 the construction of induced templates from time series data. Koçarev *et al.* derive an induced template in [106] which, by appealing to Theorem 3.3.7, we can show contains all embedded templates as subtemplates. In the literature on induced templates, it is implicit that the "physical" system may be expected to contain merely a subset of the knots and links on the induced template. Hence, the use of this induced template would appear to be of limited applicability — it contains far too much.

Remark 3.3.12 The theorems of this section can be applied to the problem of accumulations of knots on a template from §3.1. In contrast to Theorem 3.1.20, universal templates have no restrictions on the types of accumulations of knots.

Theorem 3.3.13 (Ghrist [68]) *Let* $\{K_i\}$ *be an arbitrary sequence of knot types, and let* K *be any chosen knot type. Then, on the universal template* $\mathcal{V}$, *there exists a sequence of distinct closed orbits* $\{\gamma_i\}$ *of knot type* K_i, *which accumulates onto a closed orbit* γ *of knot type* K.

This theorem sheds light on the class of *infinite* links contained in universal templates: of course, not every infinite link may live on a template, but there is no obstruction as far as accumulations of knot types goes.

Remark 3.3.14 A template contains both topological and dynamical information. By "forgetting" the topology, one reduces a template to a purely dynamical object. For example, if one takes the set of branch lines as a cross-section to the semiflow, on obtains a set of coupled, expanding, one-dimensional maps. Or, if one collapses a template along the direction transverse to the semiflow, one obtains a directed graph, which defines a subshift of finite type (*cf.* Remark 1.2.22). Theorems 3.3.2 and 3.3.7 then yield as a scholium a dynamical result:

Corollary 3.3.15 *Let* (Σ_A, σ) *be an irreducible subshift of finite type. Given any* $N \times N$ *matrix of zeros and ones,* B, *there exists a local cross section* $\Sigma' \subset \Sigma_A$ *such that the return map* r *acting on this cross section is conjugate to the subshift defined by* Σ_B.

A similar result holds with renormalizations of coupled, expanding one-dimensional maps. These dynamical results are, if not well-known, then at least provable through much simpler methods than those of this chapter. Yet, we note that the methods used in this chapter are by-and-large topological: Alexander's Theorem, braid groups, etc., are key tools. Thus, we are pleased that knot-theoretic tools can be brought to bear on a dynamical problem. In the next chapter, too, such tools will be shown to be useful in studying bifurcations of parametrized families of flows.

Chapter 4: Bifurcations

In Chapter 3 we derived general results on template knots and links. The theme was one of richness and inclusion: *every* template contains infinitely many distinct knot types; templates carrying unlinked, unknotted, untwisted orbits support infinite sequences of isotopic knots, and, most strikingly, "many" templates with mixed crossings carry *all* knots and links (and even all templates).

We now turn to issues of uniqueness and exclusion, asking how knowledge of knotting and linking data implies *restrictions* on families of periodic orbits and the bifurcations in which they are created. More specifically, in a parametised family of flows, periodic orbits appear and disappear in [often complicated] sequences of bifurcations. But for three-dimensional flows, it is the *link* of periodic orbits which undergoes bifurcations. Thus, if (1) we "dress" the periodic orbit set with knotting and linking information; and (2) we compute the topological action of bifurcations on orbits, we produce a set of bifurcation invariants derived from knot theory. This chapter will be a brief tour through several applications of this principle.

We begin with introductory remarks on local bifurcation and continuation of orbit branches and some elementary observations regarding the link structures arising in saddle node and period-multiplying (doubling and Hopf) bifurcations from closed orbits. In § 4.2 we describe a number of results on the horseshoe template $\mathcal{H}$ of Figure 2.9, the major ones being existence, non-existence and uniqueness theorems for families of torus knots of specified dynamical periods. These provide invariants which distinguish orbits, permitting us to follow them from a chaotic hyperbolic set, back to their birthplaces in parameter space, thereby determining genealogies and orders of precedence in a family of Hénon maps. Section 4.3 contains knot theoretic analogues of the self-similarity results on bifurcation sequences of the quadratic family (1.23) introduced in §1.2.3. We show how a factorisation of kneading sequences corresponds to subtemplates which are embedded copies of $\mathcal{H}$, and indicate how this may be used to determine the orbits implicated in iterated torus knots and more general cabled structures involving horseshoe knots and links. Perhaps the major interest in this work is the way in which knot invariants afford a link (pun intended) between local bifurcations and global questions.

In the final section we address global bifurcations more explicitly, describing some periodic orbit structures that appear near homoclinic orbits to saddle-type equilibria. We call attention to two types of topologically significant global bifurcations — the *gluing* bifurcations, and the bifurcations surrounding a *Shil'nikov connection*. In the case of gluing bifurcations, the issue at hand is not richness of orbits (primarily only "simple" knots appear), but of countable bifurcation sequences. In stark contrast, in the Shil'nikov scenario, we find a general case in which the universal template $\mathcal{V}$ of §3.2 is contained within the flow, thereby giving a set of (primarily dynamical) sufficient conditions under which a given ODE contains all knots and links among its periodic orbits. We close with an example of a piecewise linear ODE which satisfies the necessary hypotheses: an explicit seed from which all knots and links can be grown.

This chapter provides merely a sample of numerous results which have been obtained for specific systems. For further examples, see [87, 93, 88, 70, 118, 119, 180]. It is our hope that knotting and linking data will become increasingly useful tools in the subtle business of tracking global phenomena in the bifurcations of periodic orbits.

4.1 Local bifurcations and links

In §1.2.3 we described the three codimension-one bifurcations of maps: the saddle-node, period-doubling, and Hopf bifurcations (we also noted the symmetric pitchfork bifurcation). In the associated three-dimensional flows obtained by suspending these families, there are natural and simple implications for knotting and linking of the periodic orbits involved. Specifically, we have:

Proposition 4.1.1 *The periodic orbits implicated in a saddle-node or pitchfork bifurcation of a three dimensional flow are isotopic knots and have the same linking number with any other orbit which persists through the bifurcation point.*

Proof: We discuss the saddle-node case, as that of the pitchfork is analogous. Consider the parametrised Poincaré map on a small cross section to the flow transverse to the orbit at the bifurcation. Upon passing the parameter through the bifurcation, the fixed point becomes a pair of fixed points, one of which (say p_1) is a saddle, the other of which (say p_2) is either a source or sink.

In the case of p_2 a source and for parameter sufficiently close to the bifurcation, one branch of $W^s(p_1)$ is a small segment contained in $W^u(p_2)$ with endpoints p_1 and p_2. Hence, in the suspension of the return map, the two periodic orbits form the boundary components of an embedded annulus, and are thus isotopic.   □

Proposition 4.1.2 *The periodic orbits created in period-doubling and Hopf bifurcations are cables of the original (bifurcating) orbit.*

Proof: Following the proof of Proposition 4.1.1, one notes that the orbit of period $2T$ created in period-doubling bifurcation is the boundary of a Möbius band formed of the two-dimensional stable (or unstable) manifold associated with the eigenvalue of the Poincaré map passing through -1, whose core is the original period T orbit. As such, it is clearly a 2-cable. Similarly, since the q-periodic orbits created in a Hopf bifurcation approach those of the linearised mapping (1.21) at the bifurcation point, and this map is a rigid rotation by p/q, they are q-cables of the core period T orbit. As in Proposition 4.1.1, varying the parameter sufficiently close to the bifurcation point creates an isotopy of the orbits in phase space, which preserves cabling and linking.   □

Remark 4.1.3 Hopf bifurcations of maps resulting in periodic orbits are sometimes called *period multiplying bifurcations*, although, as noted in § 1.2.3, this name more properly refers to the special case of (two-dimensional) area-preserving maps. Here the determinant of the linearised mapping (equal to the product of the eigenvalues) is 1 and so, as one varies a parameter, the eigenvalues of an elliptic fixed point must traverse the unit circle, which they can only leave at $+1$ (a saddle-node) or -1 (period-doubling). In this case the parameter μ and cubic term (r^3) in (1.21) are identically zero, and the parameter of interest is the rotation angle φ. As φ passes each value $2\pi p/q$, a pair of q-periodic orbits of rotation number p/q generically bifurcates from the elliptic core orbit, again leading to q-cablings of the original orbit. See [122, 123], or the summary in [93] for details.

These results may be used to exclude certain global orbit branches and bifurcations in generic three dimensional flows. Following Alexander and Yorke [2] and Kent and Elgin [104], we briefly describe an example: the "noose" bifurcation.

We will need some definitions encoding twisting information for orbits in three-dimensional flows, following [2].

Definition 4.1.4 Let γ be a periodic orbit in a three-dimensional flow having associated Poincaré map with eigenvalues λ_1 and λ_2. Then γ is said to be *elliptic* if both eigenvalues have moduli satisfying one of the following conditions: either (1) the moduli are both greater than one; (2) the moduli are both less than one; or (3) the moduli are both equal to one with $\lambda_i \neq \pm 1$. When $|\lambda_1| < 1 < |\lambda_2|$, γ is an unstable saddle orbit — here there are two sub-types, depending upon the twist of the local unstable manifold $W_{loc}^u(\gamma)$, which is a two-dimensional ribbon. (See Remark 1.2.18.) If the twist is even, so $W_{loc}^u(\gamma)$ is an annulus, we call γ *hyperbolic*; if the twist is odd so that $W_{loc}^u(\gamma)$ is a Möbius band, we call γ *Möbius*.

Hyperbolic orbits have positive real eigenvalues, Möbius orbits, negative ones. All generic (non-bifurcating) periodic orbits belong to one of these three classes. Note that this terminology differs from the standard usage in dynamical systems theory.

The local bifurcation results of Propositions 4.1.1 and 4.1.2 can now be augmented. We first note that, for flows on orientable three-manifolds, the Poincaré maps are necessarily orientation preserving, implying that $\lambda_1\lambda_2 = \det(DP) > 0$. In a codimension one saddle-node, one eigenvalue $\lambda_1 = +1$, the other being bounded away from the unit circle. It follows that, of the two orbit branches created, one is elliptic and the other hyperbolic. Similar observations apply to the pitchfork bifurcation, in which either an elliptic orbit becomes hyperbolic and gives birth to two new elliptic orbits, or a hyperbolic orbit becomes elliptic and two hyperbolic orbits are born: see Figure 4.1.

In contrast, in the period-doubling bifurcation, since the critical eigenvalue is $\lambda_1 = -1$ and the associated local invariant (center) manifold has odd twist, the bifurcating (period q) orbit is Möbius on one side of the bifurcation point and elliptic on the other. The period $2q$ orbit which bifurcates off can be seen as bounding a Möbius band which is the local (weak) stable or unstable manifold of the period q core orbit. Since it goes around twice before closing, its twist is necessarily even. Thus it is either hyperbolic (if of saddle type) or elliptic (if stable, neutral or unstable): see Figure 4.1.

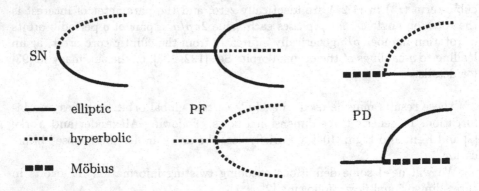

Figure 4.1: Local bifurcations of orbits, labeled as elliptic, hyperbolic, and Möbius.

Definition 4.1.5 Let γ be a hyperbolic or Möbius periodic orbit. The *self-linking number* of γ is defined as

$$slk(\gamma) = \ell k\left(\gamma', \gamma\right),$$

where γ' is a boundary component of the local unstable manifold $W^u_{loc}(\gamma)$.

Lemma 4.1.6 *Self-linking number is invariant along a continuous branch of orbits in parameter space so long as it is well-defined and the orbit path does not change type. In addition, $slk(\gamma)$ is always odd for a Möbius orbit, and, in changing from a Möbius to a hyperbolic orbit, the self linking number doubles.*

Proof: Invariance follows as before from the fact that a path of orbits in parameter space avoiding bifurcations gives an isotopy of the local unstable manifold. The remaining facts are easily shown with a picture or two, and are left as instructive exercises for the reader.   □

Note that slk may be either odd or even for hyperbolic orbits. In addition, when an orbit changes type from Möbius or hyperbolic to elliptic, self-linking is lost.

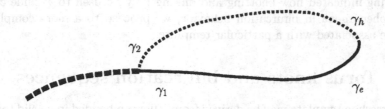

Figure 4.2: A bifurcation diagram containing a noose.

Following work of Alexander-Yorke [2] and Kent-Elgin [104], we consider the bifurcation diagram pictured in Figure 4.2: a branch of orbits loops back through a saddle node bifurcation to join itself in a period-doubling. Topologically, this requires one of the orbits born in the saddle-node to wrap around its partner as the boundary of a Möbius band. While this sort of bifurcation can generically occur in flows of dimension four and higher, there are nontrivial restrictions in dimension three:

Proposition 4.1.7 (Kent and Elgin [104]) *For a flow on $\mathbb{R}^3$ parametrised by μ, the "noose" pictured in Figure 4.2 is impossible.*

Proof: This is an exercise with linking, self-linking, and twist. The noose joins at a period-doubling point; hence the smaller period orbit γ_1 implicated in it starts either as a Möbius or an elliptic orbit, while the longer period one γ_2 is elliptic or hyperbolic. In either case, while both orbits coexist, $lk\,(\gamma_1,\gamma_2)$ is odd, and, if γ_2 is hyperbolic, $slk\,(\gamma_2)$ is even.

We further augment Proposition 4.1.1 by noting that the twist of the two-dimensional local invariant (center) manifold associated with the bifurcating eigenvalue $(+1)$ at a saddle-node is inherited by both the elliptic (γ_e) and hyperbolic (γ_h) orbits produced. Since directly after bifurcating γ_h and γ_e are "parallel" on this band, the self-linking number of the hyperbolic orbit satisfies $slk(\gamma_h) = lk(\gamma_e, \gamma_h)$. The fact that $lk(\gamma_1, \gamma_2)$ is odd near the period doubling implies that $lk(\gamma_e, \gamma_h)$ must likewise be odd, so that $slk(\gamma_h)$ is odd. But we showed that for the hyperbolic orbit, self-linking is even. $\quad\square$

Remark 4.1.8 Alexander and Yorke [2] have developed an index theory for dealing with general bifurcation diagrams. They, as well as Kent and Elgin [104], have found certain types of nooses which can live in three-dimensional flows; however, these allowable nooses involve nongeneric behaviour, such as pitchfork bifurcations, or intricate heteroclinic connections. Statements more general than that of Proposition 4.1.7 can be made which exclude these unusual cases.

Having indicated how knotting and linking may be used to exclude certain global phenomena in bifurcation behavior, we proceed to a more complicated instance associated with a particular template.

4.2   Torus knots and bifurcation sequences

The horseshoe template may be derived from a flow embedded in a solid torus, as indicated in §2.3. The underlying vector field often models a periodically forced oscillator. As such, the template's (single) branch line corresponds naturally to a global cross section in the original flow, and the number of intersections of a periodic orbit with the branch line is the dynamical as well as the symbolic period of the knot (*cf.* Remark 2.4.5). This observation prompts the following:

Definition 4.2.1 Given a (p, q) torus knot, we say it is a *resonant torus knot* if it has period q.

Recall, we may take $p < q$ without a loss of generality.

Example 4.2.2 Consider the word $\mathbf{w}^\infty = (x_1 x_2^2 x_1 x_2)^\infty$, of period five. To determine whether it is a torus knot, we draw it on the horseshoe template $\mathcal{H}$. The five points in the intersection of the knot with the branch line of $\mathcal{H}$ have addresses $\{\sigma^k(\mathbf{w}^\infty) : k = 0, 1, 2, 3, 4\}$. To determine the order in which these points are traversed as one follows the knot, we use the prescription of §1.2.3, and compute the invariant coordinates of $\mathbf{w}^\infty$ and its shifts:

Word	Invariant coordinate	Ordering
$\mathbf{w} = (x_1 x_2 x_2 x_1 x_2)^\infty$	$\theta\,(\mathbf{w}) = x_1 x_2 x_1 x_1 x_2 \ldots$	0
$\sigma\,(\mathbf{w}) = (x_2 x_2 x_1 x_2 x_1)^\infty$	$\theta(\sigma\,(\mathbf{w})) = x_2 x_1 x_1 x_2 x_2 \ldots$	2
$\sigma^2(\mathbf{w}) = (x_2 x_1 x_2 x_1 x_2)^\infty$	$\theta(\sigma^2(\mathbf{w})) = x_2 x_2 x_1 x_1 x_2 \ldots$	3
$\sigma^3(\mathbf{w}) = (x_1 x_2 x_1 x_2 x_2)^\infty$	$\theta(\sigma^3(\mathbf{w})) = x_1 x_2 x_2 x_1 x_2 \ldots$	1
$\sigma^4(\mathbf{w}) = (x_2 x_1 x_2 x_2 x_1)^\infty$	$\theta(\sigma^4(\mathbf{w})) = x_2 x_2 x_1 x_2 x_2 \ldots$	4

Drawing a simple closed curve on $\mathcal{H}$ which passes through the branch line points in the prescribed order above yields the knot corresponding to $\mathbf{w}^\infty$, as shown in Figure 4.3(a). The reader can perform Reidemeister moves to obtain Figure 4.3(b), revealing that $(x_1 x_2^2 x_1 x_2)^\infty$ is a $(2, 5)$ resonant torus knot. Similarly, it can be verified that $(x_1^2 x_2 x_1 x_2)^\infty$, also of period five, corresponds to a $(2, 3)$ torus knot, and hence is *not* resonant.

Numerous statements can be made regarding existence and uniqueness for torus knots and resonant torus knots on the horseshoe template. Before giving the first of these, which requires a lengthy proof, we state a simpler result on pairs of orbits arising in saddle node bifurcations. The key idea throughout this and the following section involves mapping sets of words to knot types, and we use extensively the ordering of points on the branch line via symbolic dynamics and kneading theory of §1.2.3. In doing so, we refer to the return map $f_{\mathcal{H}}$ induced on the branch line by the semiflow.

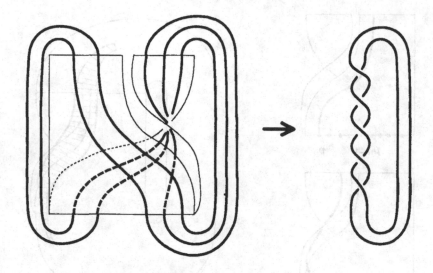

Figure 4.3: (a) The orbit $(x_1 x_2^2 x_1 x_2)^\infty$ is (b) a resonant $(2,5)$ torus knot.

Given two words corresponding to template knots, it is generally difficult to determine if the knots are isotopic. As noted earlier, this is relevant to the associated bifurcation behavior; *e.g.*, upon varying parameters in a flow, non-isotopic orbits cannot collapse onto one another in saddle-node bifurcations. However, in some cases we can perform isotopies on the template to obtain such results.

Lemma 4.2.3 *Let $\mathbf{w}^\infty$ be a periodic point on $\Sigma_\mathcal{H}$ which is minimal with respect to $\lhd$ among all its shifts. Then, if the words $\mathbf{w}x_1x_2$ and $\mathbf{w}x_2^2$ are both acyclic, then the knots on $\mathcal{H}$ corresponding to $(\mathbf{w}x_1x_2)^\infty$ and $\left(\mathbf{w}x_2^2\right)^\infty$ are isotopic.*

Proof: Let $\{p_i\}_0^n$ and $\{q_i\}_0^n$ be the points at which the orbits $(\mathbf{w}x_1x_2)^\infty$ and $\left(\mathbf{w}x_2^2\right)^\infty$ respectively intersect the branch line. These correspond symbolically to all shifts of the words $\mathbf{w}x_1x_2$ and $\mathbf{w}x_2^2$. By Proposition 1.2.47, the minimality of these words implies that $p_0 < p_k, \forall k \neq 0$ and $p_n > p_k, \forall k \neq n$, and similarly for q_0 and q_n. Since the semiflow takes p_{n-1} to p_n and p_n is maximal among the p_i points, then among all the p_i points on the left half of the branch line (that is, the strip x_1), p_{n-1} is maximal. Similarly, since the template semiflow reverses orientation on the right side (the strip x_2), then among all the q_i points on the x_2 strip, q_{n-1} is minimal. Thus, p_{n-1} and q_{n-1} lie on opposite sides of the gap in the branch line, with no other strands between them. From Figure 4.4 it is clear that one may lift the strand passing through p_{n-1} over the gap to q_{n-1}, obtaining the desired isotopy. □

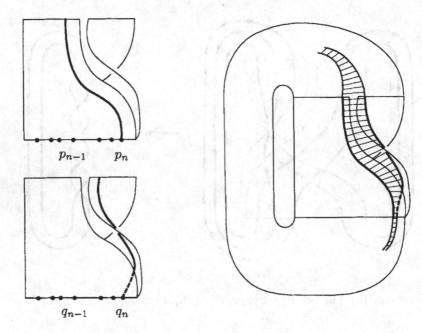

Figure 4.4: $(\mathbf{w}x_1x_2)^\infty$ is isotopic to $(\mathbf{w}x_2^2)^\infty$.

Example 4.2.4 For example, the pair $x_1x_2^2x_1x_2$ and $x_1x_2^2x_2x_2(= x_1x_2^4)$ form such a minimal acyclic pair. As noted above, $(x_1x_2^2x_1x_2)^\infty$ is a $(2,5)$ torus knot; thus, so is $(x_1x_2^4)^\infty$.

Definition 4.2.5 Two minimal acyclic words of the form $\mathbf{w}x_1x_2$ and $\mathbf{w}x_2^2$ are called a *bifurcation pair*. These two words have differing x_2-parities: we denote that with even x_2-parity *male* and that with odd x_2-parity *female*.

Remark 4.2.6 The reason for the terminology of Definition 4.2.5 is as follows: recall that the return map for $\mathcal{H}$ induced by the branch line can be considered as a member of the quadratic family of maps (§1.2.3). If we then regard horseshoe knots as periodic orbits created as one passes through a sequence of quadratic maps, Proposition 1.2.48 implies that the male-female pair from Definition 4.2.5 is created simultaneously in a saddle-node bifurcation. In this and the following section, we will freely pass from thinking of finite words in $\{x_1, x_2\}$ as horseshoe knots or as periodic points in the quadratic family. These "genders" reflect the role played by the knots in orbit genealogies, to be detailed in §4.3.

Lemma 4.2.3 does not imply that all knots come in isotopic pairs. Take, for example, the period four orbit $(x_1x_2^3)^\infty$, whose bifurcation partner would be $(x_1x_2x_1x_2)^\infty$: a cyclic extension of the period two word x_1x_2. Evidently $x_1x_2^3$ has no partner. Such a "pseudo-pair" is related to a period-doubling bifurcation within the quadratic family, in analogy to the saddle-node pairs of Remark 4.2.6: cf. §4.3.

Other results similar to Lemma 4.2.3 are possible. The following is a corollary to Proposition 3.1.19, easily proved in this special case by removing an "x_1-loop" via the first Reidemeister move:

Corollary 4.2.7 *If* w *is minimal, then the knots corresponding to* $\left(x_1^k \text{w}\right)^\infty$ *are isotopic for all* $k \geq 0$.

Before stating the main theorem of this section, we need a further result which enables us to easily determine the braid index for a class of positive braids (recall Definition 1.1.23).

Theorem 4.2.8 (Franks and Williams [58]) *For a positive braid on* p *strands containing a full twist on* p *strands, the braid index is* p.

The proof of Theorem 4.2.8 uses Jones polynomials and is beyond the scope of this book.

4.2.1   Horseshoe torus knots

Theorem 4.2.9 (Holmes and Williams [93]) *Among the* (p,q) *torus knots on* $\mathcal{H}$, *there are:*

1. *exactly two resonant torus knots for each* $q > 2p$, *and infinitely many nonresonant torus knots of arbitrarily large period;*

2. *no resonant torus knots for* $q < 2p$;

3. *no torus knots at all for* $q < 3p/2$.

In addition to supplying a specific instance of an infinite collection of distinct knot types on $\mathcal{H}$ (which we expect from Theorem 3.1.15), this theorem reveals that the *resonant* torus knots are surprisingly sparse. It also suggests that the additional positive half-twist on $\mathcal{H}$ makes it more "rigid" than the Lorenz template $\mathcal{L}(0,0)$, which contains *all* torus knots by Theorem 2.3.3.

Outline of proof: To prove the existence of the resonant torus pair for $q > 2p$, we extract a subset S from the horseshoe template (S is *not* a subtemplate as there are no branches). In Figure 4.5 we show $\mathcal{H}$ without its ends identified. We remove portions on the edges of the x_1-branch and the center of the x_2-branch (a neighborhood of the orbit x_2), yielding three strips which can be laid on a cylinder. Identifying the ends of the cylinder, we have a torus T^2 on which S lies.

A (p,q) resonant torus knot has q strands traveling p times meridionally about T^2. We construct one by placing p strands on each of the two x_2-strips and $q - 2p$ strands on the x_1-strip of S (see Figure 4.5). The partner is obtained

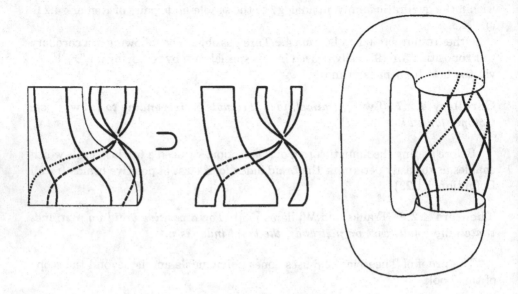

Figure 4.5: The resonant torus knots on $\mathcal{H}$.

by reversing the isotopy in the proof of Lemma 4.2.3, lifting the leftmost x_2-strand over to form the rightmost x_1-strand.

To specify the words for this pair, write a string of x_1's of length $q - 2p$ followed by two strings of x_2's, each of length p. The first word is produced by counting forward in multiples of p mod q: beginning at the first x_1 and recording the appropriate letter, each time advancing p letters and "wrapping around" where necessary, regarding the sequence as periodically extended. The partner derives from Lemma 4.2.3, on changing the penultimate letter from x_2 to x_1. The first x_2 in the first group of p x_2's is the ambivalent term for the pair, denoted below by x_*. Note that these words have x_1's and x_2's distributed in the most uniform manner possible, subject to the required relative number $2p/q$ or $(2p-1)/q$ of x_2's. Hence they are sometimes called *evenly distributed* words [91].

Example 4.2.10 To determine the (3,11) resonant torus pair, write out the prescribed string of x_1's and x_2's:

$$\overbrace{x_1\ x_1\ x_1\ x_1\ x_1}^{11 - 2 \cdot 3} \mid \overbrace{x_2\ x_2\ x_2}^{3} \mid \overbrace{x_2\ x_2\ x_2}^{3}, \tag{4.1}$$

then, counting terms mod 11, one gets $x_1\ x_1\ x_2\ x_2\ x_1\ x_1\ x_2\ x_2\ x_1\ x_2\ x_2$. Hence, the resonant torus knot pair is given by $(x_1^2 x_2^2)^2 x_1 x_* x_2$.

Given any pair of resonant torus knots, Corollary 4.2.7 immediately yields infinitely many more isotopic but nonresonant ones, of periods $q + 1, q + 2, \ldots$.

The uniqueness proof is more complicated. The idea, due to Williams, is to rearrange orbits on the template $\mathcal{H}$ in minimal or "well-disposed" braid form and use braid index and genus invariants together with dynamical period. The details appear in [93]; here we sketch only principal ideas.

To apply Theorems 1.1.18 and 4.2.8 in computing genera and braid indices, we must transform orbits on $\mathcal{H}$ into the appropriate form:

Proposition 4.2.11 *With the exception of the orbits x_1^∞ and x_2^∞, every orbit on $\mathcal{H}$ may be arranged as a positive braid having a full twist.*

Proof: We first perform a DA-splitting on the x_2^∞ orbit, creating an isolated source (what was x_2^∞) linking the DA-modified template, which has a new boundary component corresponding to $\left(x_2^2\right)^\infty$. This DA modification affects only the x_2^∞ orbit (which is to become the braid axis) and the new boundary component: all other orbits are unchanged.

After removing the braid axis and propagating the branch line gaps back, loops are transformed into full twists, via the belt trick, as illustrated in Figures The template is thereby transformed to a positive braid with the exception of a loop at the top, corresponding to the x_1-strip of $\mathcal{H}$. For any given link with total number of consecutive x_1's bounded, we may split the x_1 branch line repeatedly as before and pull each curl out via the belt trick, producing a subtemplate of $\mathcal{H}$ containing the link as a positive braid with (at least) one and one-half full twists: more than sufficient for application of Theorem 4.2.8. $\square$

Equipped with this "normal form" for $\mathcal{H}$ and given a knot with periodic word $\mathbf{w} = x_1^{a_1} x_2^{b_1} x_1^{a_2} x_2^{b_2} \dots x_1^{a_k} x_2^{b_k}$, we define *syllables* to be of the form $x_1^n x_2, x_1^n x_2^2$, or x_2^2, for arbitrary $n > 0$. Figure and x_2, each word has a unique syllabic decomposition and each syllable corresponds to a single strand on the minimal template. Thus, via Theorem 4.2.8 we have:

Proposition 4.2.12 *The braid index of a horseshoe knot equals the number of syllables in its word $\mathbf{w}$.*

Example 4.2.13 The knot $x_1^2 x_2 x_1 x_2 x_1^3 x_2^3$ has braid index four via the decomposition $(x_1^2 x_2)(x_1 x_2)(x_1^3 x_2)(x_2^2)$.

To prove uniqueness of resonant torus knots, one shows that, among *all* braids on p-strands which cross the branch line q times, including multicomponent links, the members of the (p, q) torus knot pair alone maximize the genus. This is done via Theorem 1.1.18 by maximizing the crossing number c of q-period p-braids on the positive braid template, in a manner similar to the proof of Corollary 3.1.11. The calculations are presented in full in [93]. This completes the proof of part (1) in Theorem 4.2.9. The proof of (2) follows from the same calculations performed for part (1).

The proof of part (3) is simpler, and provides a nice example of the use of knot invariants. As one can verify, the braid word for a full twist on n-strands

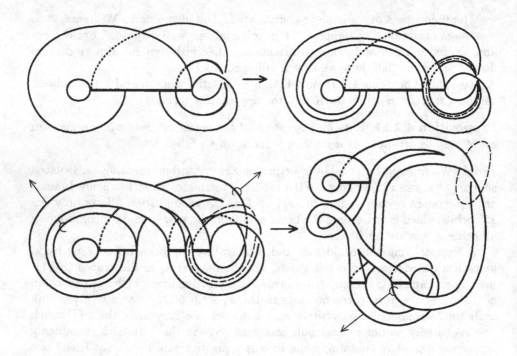

Figure 4.6: Moves to obtain the minimal template (1).

is

$$(\sigma_1\sigma_2\ldots\sigma_{n-1})^n, \tag{4.2}$$

and thus, a full twist contains $(n-1)(n)$ crossings. The minimal braid template includes three half-twists and so any braid β with braid index $b(\beta) = p$ must have crossing number $c(\beta) \geq \frac{3}{2}(p-1)(p)$. Thus, applying Theorem 2.2.4 to a (one component) knot, we have:

$$2g(\beta) \geq \frac{3}{2}(p-1)(p) - p + 1,$$

or

$$g(\beta) \geq \frac{(p-1)(\frac{3}{2}p-1)}{2}.$$

But, recalling from §1.1.4 that the genus of a (p,q) torus knot is $\frac{1}{2}(p-1)(q-1)$, we conclude that, in order to satisfy

$$\frac{(p-1)(q-1)}{2} \geq \frac{(p-1)(\frac{3}{2}p-1)}{2},$$

we must have $q > 3p/2$.   □

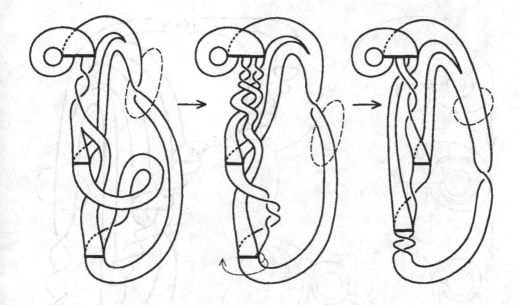

Figure 4.7: Moves to obtain the minimal template (2).

4.2.2 Bifurcation reversal in the Hénon map

Theorem 4.2.9 implies that for each pair of relatively prime positive integers (p, q) with $q > 2p$, the flow in the suspension of the horseshoe map has a unique pair of resonant (p, q) torus knots. We will now relate this information to bifurcation sequences involving such orbit pairs in the Hénon map (2.12). As noted in §2.3.2, for $\mu > \frac{1}{4}(5 + 2\sqrt{5})(1 + \epsilon^2)$, the map $F_{\mu,\epsilon}$ has a horseshoe and so, suspending this family as in Figure 2.9, we have the resonant torus knots described above.

For the case $\epsilon = 1$, the map $F_{\mu,\epsilon}$ becomes

$$F_{\mu,1} : \begin{cases} u \mapsto v \\ v \mapsto -u + \mu - v^2 \end{cases}, \tag{4.3}$$

an area-preserving family. Elementary calculations show that, at $\mu = -1$, $F_{\mu,1}$ undergoes a saddle-node bifurcation, creating an elliptic fixed point which persists in the interval $\mu \in (-1, 3)$. Increasing μ from -1 to 3, each member of the eigenvalue pair travels around the unit circle monotonically, taking on all values $(e^{2\pi i\phi}, e^{-2\pi i\phi})$ beginning at $(+1, +1)$ for $\mu = -1$ and ending at $(-1, -1)$ for $\mu = 3$. Using normal forms, Holmes and Williams [93] show that as the eigenvalues of $DF_{\mu,1}$ pass through each pair $(e^{2\pi ip/q}, e^{-2\pi ip/q})$ for p, q relatively prime, $q > 2p$, and $q \geq 5$, the map $F_{\mu,1}$ undergoes a generic resonant area-preserving Hopf bifurcation, creating a pair of isotopic orbits. In the natural suspension of the map, one uses Proposition 4.1.2 to show that this pair is a (p, q) resonant torus knot pair. The order in which the eigenvalues pass through the points $(e^{2\pi ip/q}, e^{-2\pi ip/q})$ determines the bifurcation sequence. By a com-

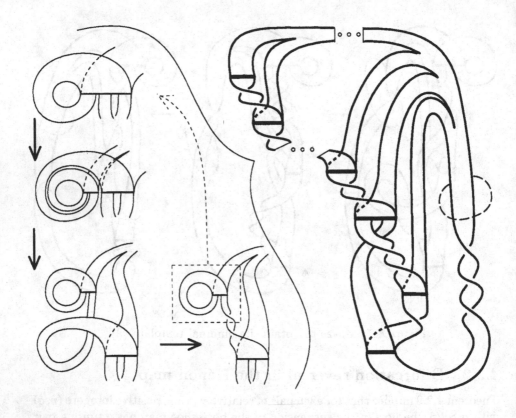

Figure 4.8: Moves to obtain the minimal template (3).

plicated argument involving symmetry properties of the map $F_{\mu,1}$ and linking data on the q-cables themselves, one shows that the resonant bifurcation pair lie on a continuous branch of resonant torus knots which can be followed from the bifurcation point to μ arbitrarily large, thus identifying them with the unique resonant pair and enabling one to employ the uniqueness part of Theorem 4.2.9 to arrive at the following:

Proposition 4.2.14 *Given the sequence of pairs of relatively prime positive integers $\{(p_i,q_i)\}_{-\infty}^{+\infty}$ with $q > 2p$ and $q \geq 5$ ordered via $i < j$ if and only if $p_i/q_i < p_j/q_j$, let μ_i^1 be the μ-value at which the natural suspension of the map $F_{\mu,1}$ creates the unique pair of (p_i,q_i) resonant torus knots. Then $i < j$ if and only if $\mu_i^1 < \mu_j^1$.*

For the case $\epsilon = 0$, the map $F_{\mu,\epsilon}$ becomes

$$F_{\mu,0} : \quad \begin{cases} u \mapsto v \\ v \mapsto \mu - v^2 \end{cases}, \tag{4.4}$$

the dynamics of which immediately collapse to those of the one-dimensional quadratic map $f_\mu : x \mapsto \mu - x_1^2$ described in §1.2.3. Kneading theory provides a complete ordering of the bifurcations of f_μ via the kneading invariants $\nu(\mathbf{w})$. Here the continuation with increasing μ of orbits once created is assured by the monotonicity of the kneading invariant. One uses the algorithm given as Example 4.2.10 of §4.2.1 to construct the words corresponding to such (p,q) resonant torus partners. Computing the associated kneading invariants (via (1.24)), Proposition 1.2.48 allows us to order these resonant torus pair bifurcations. This yields:

Proposition 4.2.15 *Given the sequence of pairs of relatively prime positive integers* $\{(p_i, q_i)\}_{-\infty}^{+\infty}$ *with* $q \geq 2p$ *ordered via* $i < j$ *if and only if* $p_i/q_i < p_j/q_j$, *let* μ_i^0 *be the* μ-*value at which the natural suspension of the map* $F_{\mu,0}$ *creates the unique pair of* (p_i, q_i) *resonant torus knots. Then* $i < j$ *if and only if* $\mu_i^0 > \mu_j^0$.

We note that the kneading theory behind Proposition 4.2.15 applies to *any* unimodal function of v in place of $\mu - v^2$ in $F_{\mu,0}$. Thus, the conclusion holds for a far wider class of mappings than the Hénon family.

These propositions together imply the following remarkable result [93, 87]:

Theorem 4.2.16 (Holmes and Williams [93]) *In the bifurcation diagram of the map* $F_{\mu,\epsilon}$, *infinitely many saddle-node bifurcation curves cross one another on the* (μ, ϵ) *parameter plane between* $\epsilon = 0$ *and* $\epsilon = 1$. *In particular, each resonant torus bifurcation sequence for the area-preserving case* ($\epsilon = 1$) *is exactly reversed in the one-dimensional case* ($\epsilon = 0$).

Thus, fixing $\epsilon \in [0,1]$, and increasing μ, we obtain infinitely many different bifurcation sequences leading to a horseshoe: loosely speaking – infinitely many routes to chaos. However, this behavior does not imply similar reversals for other orbits. For example, the (2,3) *non-resonant* torus knots of periods 4,5,6... do not reverse their order in this way; instead, as an accumulating family of the type described in Theorem 3.1.20, their bifurcation curves are all "parallel:" *cf.* Holmes and Whitley [92].

4.3   Self-similarity and horseshoe cables

Given the correspondence between knotted orbits on the horseshoe template and bifurcations of the one-dimensional quadratic family touched on in §4.2, we now explore this latter family of maps in greater detail.

Denote by f_μ the map which takes x to $\mu - x_1^2$, where f_μ acts on the interval $I(\mu) = \left[-\frac{1}{2} - \sqrt{\mu + \frac{1}{4}}, \frac{1}{2} + \sqrt{\mu + \frac{1}{4}} \right]$ (this interval grows as μ ranges over $[-\frac{1}{4}, 2]$). The bifurcation set of this map has a remarkable self-similar structure: given any positive integer M, there exists at least one subset J, of the phase-parameter space for which

$$f_\mu^M \big|_J \approx f_\mu, \tag{4.5}$$

where $\approx$ denotes conjugacy. Figure 4.9 illustrates the case $M = 3$: f_μ^3 restricted to a subinterval $[\alpha, \beta]$ has the same bifurcation sequence on some μ-subinterval as does f_μ on $[-\frac{1}{4}, 2]$). This is the basis for a renormalization group theory (see [95, 96]) which shows that bifurcation sequences are nested within themselves. The simplest such nesting leads to the well-known period-doubling cascades studied metrically by Feigenbaum and others (see [41, 199]).

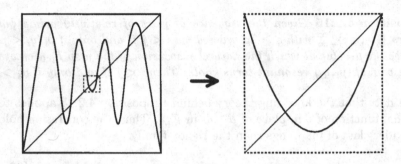

Figure 4.9: f_μ^3 and a magnification.

Orbits in the quadratic map are created in a very specific order, governed by the kneading invariants (Proposition 1.2.48). As we have seen in §4.2.2, a horseshoe may be "built" through a variety of distinct paths; nevertheless, by taking the branch line of $\mathcal{H}$ as a Poincaré section for the semiflow, we recover the "full" quadratic map as a return map. Thus, as per Remark 4.2.6, we may speak of two horseshoe knots being a saddle-node pair, based on the corresponding theory for the one-dimensional return map.

In this section, we explore the implications of the bifurcation structures within f_μ on knot and link types and on subtemplate structures within the template $\mathcal{H}$. We first outline an extension to the simple kneading theory introduced in §1.2.3, and use it to show how certain classes of words correspond to knots inhabiting subtemplates of $\mathcal{H}$. This material is drawn from [88], in which the idea of subtemplates first appeared, but the proof of the main result (Theorem 4.3.8) is reformulated and simplified in terms of the template inflations introduced in Chapter 2.

4.3.1   Kneading factorization and subtemplates

The kneading invariant introduced in §1.2.3 provides a convenient symbolic tool for analyzing iterated structures on the template. For the horseshoe template, the kneading invariant $\nu(\mathbf{a}^\infty)$ of a periodic orbit $\mathbf{a}^\infty$ is a sequence given by (1.24) which, via Proposition 1.2.48, allows one to order the μ-values at which the orbits appear in bifurcations of the one-dimensional map f_μ. In cases where $\nu(\mathbf{a})$ is periodic, we refer to it by the periodically repeated unit, with the superscript ∞ dropped.

We now describe a factorization of such kneading sequences.

Definition 4.3.1 For $\mathbf{w}$ an acyclic minimal word $\mathbf{w} = w_1 w_2 \ldots w_k$ and $\mathbf{v}$ any word $\mathbf{v} = v_1 v_2 \ldots$ (not necessarily finite), define $\mathbf{w} * \mathbf{v}$ to be the sequence of concatenated words

$$\mathbf{w} * \mathbf{v} = \mathbf{w}^{v_1} \mathbf{w}^{v_2} \mathbf{w}^{v_3} \ldots \tag{4.6}$$

where

$$\mathbf{w}^{x_1} = \mathbf{w} = w_1 w_2 \ldots w_k$$

and

$$\mathbf{w}^{x_2} = \hat{\mathbf{w}} = \hat{w}_1 \hat{w}_2 \ldots \hat{w}_k.$$

Recall from §1.2.3 that $\hat{x_1} = x_2$ and *vice versa*.

Example 4.3.2

$$
\begin{aligned}
x_1 x_2 * x_1 x_1 x_2 &= x_1 x_2\, x_1 x_2\, x_2 x_1 \\
&= x_1 x_2 x_1 x_2^2 x_1 \\
x_1 x_2 * x_1 x_2 x_2 &= x_1 x_2\, x_2 x_1\, x_2 x_1 \\
&= x_1 x_2^2 x_1 x_2 x_1 \\
x_1 x_1 x_2 * x_1 x_2 x_1 x_1 x_2 \ldots &= x_1 x_1 x_2\, x_2 x_2 x_1\, x_1 x_1 x_2\, x_1 x_1 x_2\, x_2 x_2 x_1 \ldots
\end{aligned}
$$

Any kneading invariant ν which can be expressed as a *-product of two or more nonempty words is said to be *-*factorizable*, otherwise it is *-*prime*. The *-factorization is particularly useful in describing period multiplying bifurcations. For example, in the period-doubling bifurcation of a period-k orbit with periodic kneading invariant $\mathbf{w}$, the new orbit of period $2k$ has kneading invariant $\mathbf{w} * x_1 x_2$. The *-products can be iterated to form longer, more complicated factorizations.

The self-similarity for the quadratic map f_μ in (4.5) is naturally expressed in terms of kneading sequences and *-factorization (see [88]):

Lemma 4.3.3 *Let* $\mathbf{u}, \mathbf{v}$ *and* $\mathbf{w}$ *be kneading invariants, where* $\mathbf{w}$ *is finite. Then* $\mathbf{w} * \mathbf{u} \lhd \mathbf{w} * \mathbf{v}$ *if and only if* $\mathbf{u} \lhd \mathbf{v}$.

Proof: By Equation (4.6),

$$
\begin{aligned}
\mathbf{w} * \mathbf{u} &= \mathbf{w}^{u_1} \mathbf{w}^{u_2} \ldots \\
\mathbf{w} * \mathbf{v} &= \mathbf{w}^{v_1} \mathbf{w}^{v_2} \ldots
\end{aligned}
$$

Let K denote the index of the first letter at which $\mathbf{u}$ and $\mathbf{v}$ differ; hence, $u_K = x_1, v_K = x_2$. Since $\mathbf{w}$ is a kneading invariant, it follows from (1.24) that $w_1 = x_1$. Thus, $\mathbf{w}^{u_K} \lhd \mathbf{w}^{v_K}$ and $\mathbf{w} * \mathbf{u} \lhd \mathbf{w} * \mathbf{v}$. Reversing the argument yields the lemma. □

Remark 4.3.4 In conjunction with Proposition 1.2.48 and Equation (1.24), Lemma 4.3.3 implies the self-similarity in the bifurcation structure stated in Equation (4.5). Increasing μ creates periodic orbits in the order of increasing ν

from x_1^∞ to $(x_1x_2)^\infty$, to $(x_1x_2)*(x_1x_2)$, etc., *ad infinitum*. For any finite *-prime kneading words $\mathbf{u} \lhd \mathbf{v}$, all kneading sequences of the form $\mathbf{u}*\mathbf{w}$, for all $\mathbf{w}$, must preceed $\mathbf{v}$; hence, the entire bifurcation sequence of f_μ is "embedded" within itself, so that f_μ^M restricted to some subinterval in μ undergoes the "same" sequence as f_μ itself.

Recall from Definition 4.2.5 that *male* knots have even x_2-parity and *female* knots, odd x_2-parity. The kneading theory for unimodal maps implies that males are created in saddle-node bifurcations and females in either saddle-nodes (along with males) or, partnerless, in period-doubling bifurcations. Directly after either such bifurcation, both orbits implicated in it share the same symbol sequence. After the saddle node, that destined to become female changes gender via one point on it crossing the critical point c; the male's sequence remains as it began, consistent with a positive eigenvalue. After a $q \to 2q$ period doubling, the doubled orbit, whose sequence, regarded as $2q$-periodic, starts out male, similarly changes gender by losing or gaining an x_2 as a point of it passes c. (Recall that the eigenvalue of the (iterated) maps are respectively 1 and -1 in these bifurcations.) These observations imply the following (for details see [88]):

Lemma 4.3.5 *Let* $\mathbf{w}$ *be a* q-*periodic kneading invariant. Corresponding to* $\mathbf{w}$ *and* $\mathbf{w} * x_1x_2$, *there exist two horseshoe periodic orbits,* $(\mathbf{a}')^\infty$ *and* $(\mathbf{a})^\infty \in \Sigma_\mathcal{H}$, *such that:*

1. $\mathbf{w} = \nu((\mathbf{a})^\infty) = \nu((\mathbf{a}')^\infty)$;

2. *if* $\mathbf{w} \neq \mathbf{u} * x_1x_2$ *for any kneading invariant* $\mathbf{u}$, *then* $(\mathbf{a}')^\infty$ *and* $(\mathbf{a})^\infty$ *are a male-female pair of isotopic period-q orbits created in a saddle-node bifurcation;*

3. *if* $\mathbf{w} = \mathbf{u} * x_1x_2$ *for some kneading invariant* $\mathbf{u}$, *then* $(\mathbf{a}')^\infty$ *and* $(\mathbf{a})^\infty$ *are both female knots implicated in a period-doubling bifurcation and having respective periods q and $2q$.*

Definition 4.3.6 Let $\{\mathbf{w}_i\}_1^n$ denote a collection of q_i-periodic kneading invariants for some $n > 1$, and $\mathcal{W} = \mathbf{w}_1 * \mathbf{w}_2 * \cdots * \mathbf{w}_n$ be the $Q = \prod_{i=1}^n q_i$-periodic kneading invariant formed by iterated *-multiplication. A periodic horseshoe orbit $(\mathbf{a})^\infty$ having kneading invariant $\nu((\mathbf{a})^\infty) = \mathcal{W}$ is called an *iterated horseshoe knot* with *defining sequence* $\mathcal{W}$.

The factorization of kneading invariants becomes the dynamical backbone for an elegant interpretation of self-similarity in the bifurcations of the horseshoe. The topological analogue of the *-factorization is a generalization of the satellite-companion construction for knots (Definition 1.1.10):

Definition 4.3.7 Let $\mathcal{T}$ be a template braided within a standardly embedded solid torus $V = D^2 \times S^1$, and let K be a knot (in a different copy of S^3) with tubular neighborhood $N(K)$ homeomorphic to V via $h : V \to N(K)$. Then the template given by $h(\mathcal{T})$ is a *satellite* of $\mathcal{T}$ with *companion* K.

Theorem 4.3.8 (*cf.* Holmes [88]) *Let* $\mathcal{W} = \mathbf{w}_1 * \mathbf{w}_2 * \cdots * \mathbf{w}_n$ *be a periodic kneading invariant which does not factor as* $\mathbf{u} * (x_1 x_2)$ *for any kneading invariant* $\mathbf{u}$. *Also, denote by* $(\mathbf{a}')^\infty$ *and* $(\mathbf{a})^\infty \in \Sigma_{\mathcal{H}}$ *the male-female pair of knots associated to* $\mathcal{W}$ *via Lemma 4.3.5. Then, all the iterated horseshoe knots of the form* $\mathcal{W} * \mathbf{v}$ *coincide with the closed orbits on a particular subtemplate* $\mathcal{H}_{\mathcal{W}} \subset \mathcal{H}$ *which is the satellite of either the standard horseshoe template* $\mathcal{H}$ *or the "twisted" horseshoe template* $\tilde{\mathcal{H}}$ *(pictured in Figure 4.10), with the knot corresponding to* $(\mathbf{a})^\infty$ *as companion.*

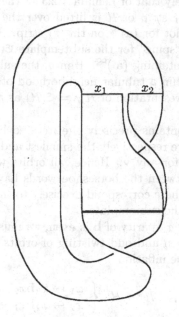

Figure 4.10: The "twisted" horseshoe template $\tilde{\mathcal{H}}$.

Proof: Let $(\mathbf{a}')^\infty$ be the (Q-periodic) itinerary of the male horseshoe knot having kneading invariant $\nu((\mathbf{a}')^\infty) = \mathcal{W}$ and let $(\mathbf{a})^\infty$ correspond to the female knot having kneading invariant $\nu((\mathbf{a})^\infty) = \mathcal{W} * x_1 x_2$ as per Lemma 4.3.5. Denote by $\mathbf{b}$ be the subword $a_1 a_2 \ldots a_{Q-2}$ of $\mathbf{a}$ (or, equivalently, $\mathbf{a}'$).

Assume first that $\mathbf{b}$ has odd x_2-parity; then, consider the inflation

$$\mathfrak{E}_{\mathcal{W}} : \mathcal{H} \hookrightarrow \mathcal{H} \quad \left\{ \begin{array}{l} x_1 \mapsto x_2 \mathbf{b} x_1 \\ x_2 \mapsto x_2 \mathbf{b} x_2 \end{array} \right. \tag{4.7}$$

The image of this map is a template since $\mathfrak{E}$ preserves the twist orientation of

x_1^{∞} and x_2^{∞}, and since the image of the branch segments,

$$\beta_1(\mathfrak{E}_W(\mathcal{H})) = \mathfrak{E}_W([x_1^{\infty}, x_1 x_2 x_1^{\infty}])$$
$$= [((x_2 \mathbf{b} x_1))^{\infty}, x_2 \mathbf{b} x_1 x_2 \mathbf{b} x_2 (x_2 \mathbf{b} x_1)^{\infty}];$$

$$\beta_2(\mathfrak{E}_W(\mathcal{H})) = \mathfrak{E}_W([(x_2)^2 x_1^{\infty}, x_2 x_1^{\infty}])$$
$$= [(x_2 \mathbf{b} x_2)^2 (x_2 \mathbf{b} x_1)^{\infty}, x_2 \mathbf{b} x_2 (x_2 \mathbf{b} x_1)^{\infty}].$$

$$(4.8)$$

is a set of nonoverlapping intervals on the branch line (this may be verified using the $\triangleleft$-ordering and the fact that $\mathbf{b}$ is of odd x_2-parity).

By Lemma 4.3.5, the knots corresponding to $(\mathbf{a})^{\infty}$ and $(\mathbf{a}')^{\infty}$ on $\mathcal{H}$ are isotopic: the isotopy is merely that of Lemma 4.2.3 — the rightmost strand of the knot for $(\mathbf{a}')^{\infty}$ on the x_1 strip of $\mathcal{H}$ is lifted over the branch line gap to the leftmost strand of the knot for $(\mathbf{a})^{\infty}$ on the x_2 strip. Since the periodic orbits $(\mathbf{a})^{\infty}$ and $(\mathbf{a}')^{\infty}$ form a "spine" for the subtemplate $\mathfrak{E}(\mathcal{H})$, the isotopy may be extended to the strip containing $(\mathbf{a}')^{\infty}$. Hence, the subtemplate $\mathfrak{E}(\mathcal{H})$ may be isotoped in S^3 to lie within a tubular neighborhood of the knot corresponding to $(\mathbf{a})^{\infty}$. This yields a presentation of $\mathcal{H}_N = \mathfrak{E}(\mathcal{H})$ as a satellite template with companion $(\mathbf{a})^{\infty}$.

To show that $\mathcal{H}_N$ contains precisely the iterated horseshoe knots, observe that W and $W * x_1 x_2^{\infty}$ are respectively the smallest and largest kneading invariants of the form $W * \mathbf{v}$ for any $\mathbf{v}$. Hence, all orbits with kneading invariants of this form must lie between the horseshoe words having kneading invariants W and $W * x_1 x_2^{\infty}$. But these correspond precisely to the boundary components $\mathfrak{E}(x_1^{\infty})$ and $\mathfrak{E}(x_2 x_1^{\infty})$ of the subtemplate.

In the case where the x_2-parity of $\mathbf{b}$ is even, we must modify the inflation $\mathfrak{E}$ to one which respects even and odd twisting of orbits. An analogous proof to that above, applied to the inflation

$$\tilde{\mathfrak{E}}_W : \tilde{\mathcal{H}} \hookrightarrow \tilde{\mathcal{H}} \quad \begin{cases} x_1 \mapsto x_2 \mathbf{b} x_2 \\ x_2 \mapsto x_2 \mathbf{b} x_1 \end{cases}, \qquad (4.9)$$

shows that the subtemplate containing the iterated horseshoe knots is a satellite of the "twisted" horseshoe template $\tilde{\mathcal{H}}$.   □

Example 4.3.9 Let $W = x_1 x_2^3 x_1$, so that $\mathbf{a}' = x_1 (x_1 x_2)^2$ and $\mathbf{a} = x_1^2 x_2^3$. The inflation is:

$$\mathfrak{E}_{x_1 x_2^3 x_1} : \mathcal{H} \hookrightarrow \mathcal{H} \quad \begin{cases} x_1 \mapsto x_2 (x_1^2 x_2) x_1 \\ x_2 \mapsto x_2 (x_1^2 x_2) x_2 \end{cases}.$$

Figure 4.11 shows that, after an isotopy, $\mathcal{H}_N$ is a satellite of $\mathcal{H}$ with companion a trefoil, having an additional four full twists.

Similarly, the net twisting for $\mathcal{H}_N$ with $W = x_1 x_2^2 x_1 x_2$ is odd. Recalling Example 4.2.2, the reader may like to check that this subtemplate is a satellite of the twisted horseshoe $\tilde{\mathcal{H}}$ with companion a $(2, 5)$ torus knot, having four and one-half full twists.

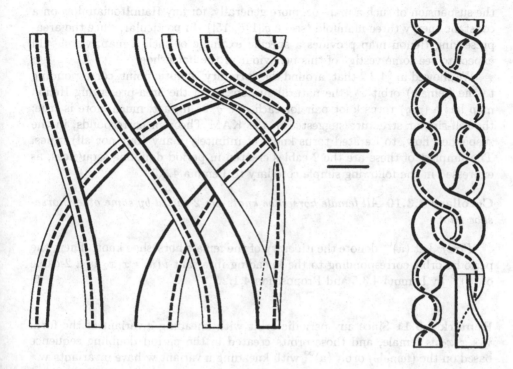

Figure 4.11: the subtemplate for $\mathcal{W} = x_1 x_2^3 x_1$. $(\mathbf{a}')^\infty$ is shown solid, $(\mathbf{a})^\infty$ dashed.

In this sense, the male-female pair $(\mathbf{a}')^\infty$, $(\mathbf{a})^\infty$ give rise to a family of iterated horseshoe knots which remain close to them in that they lie on the subtemplate $\mathcal{H}_\mathcal{N}$. We refer to $(\mathbf{a}')^\infty$ and $(\mathbf{a})^\infty$ as the *father* and *mother knots* respectively; the iterated knots are their *children*. From Theorem 4.3.8, we observe that $(\mathbf{a})^\infty$ can be viewed as the core of an embedded torus, with $(\mathbf{a}')^\infty$ on its boundary and all subsequent children following $(\mathbf{a})^\infty$ without doubling back. Hence, iterated horseshoe knots are examples of the generalized cablings discussed in §1.1.2.

4.3.2 Nested periodic orbits and iterated torus knots

The self-similarity in the bifurcation structure of the quadratic family is not the only example of dynamical self-similarity. A very important and well-known class of examples is given in the *KAM theory* for elliptic fixed points of an area-preserving diffeomorphism [122, 123]. Let $F : R^2 \to R^2$ and $DF(\xi)$ have eigenvalues $\lambda, \overline{\lambda} = e^{\pm 2\pi i \phi}$ with $\phi \in (0, \frac{1}{2})$. Generically F is a perturbed twist map with rings of alternating elliptic and hyperbolic points arranged in a self-similar fashion. These families of periodic points are separated by invariant "KAM curves," which form a set of positive Lebesgue measure; see [8].

There is much to be said concerning the knotting and linking of orbits in the suspension of such a map, or, more generally, for any Hamiltonian flow on a constant-energy three-manifold (see, *e.g.*[116, 15]). In particular, since the area-preserving Hénon map provides a specific example of such a map, we should expect to see some vestige of this behaviour in the horseshoe.

We showed in §4.2.2 that, around the "primary" elliptic point, corresponding to the (female) orbit x_2, the natural suspension of the area-preserving Hénon map has a (p, q) torus knot pair for each $p < q/2$. In fact, much more is true: the self-similar structure suggested in the KAM Theorem corresponds, in the suspended flow, to iterated torus knots of infinitely many (but not all) types. The simplest of these are the 2-cables created in period doubling sequences, as expressed in the following simple corollary to Lemma 4.3.5:

Corollary 4.3.10 *All female horseshoe knots are 2-cabled by some other horseshoe knot.*

Proof: Let $(\mathbf{a})^\infty$ denote the itinerary of the female horseshoe knot. Then, the periodic orbit corresponding to the kneading invariant $\nu(\mathbf{a}) * x_1 x_2$ is a 2-cable of $(\mathbf{a})^\infty$ by Lemma 4.3.5 and Proposition 4.1.2.   □

Remark 4.3.11 Since any periodic orbit with kneading invariant of the form $\mathbf{w} * x_1 x_2$ is female, and those orbits created in the period doubling sequence based on the (female) orbit $(\mathbf{a})^\infty$ with kneading invariant $\mathbf{w}$ have invariants $\mathbf{w} * x_1 x_2$, $\mathbf{w}*x_1 x_2 *x_1 x_2$, $\ldots$, any finite part of every period doubling sequence forms an iterated 2-cable of $(\mathbf{a})^\infty$. Formulae describing crossing and linking numbers of such structures may be derived. For example, see [88] for a presentation of the period-doubling cascade results of Yorke and Alligood [198, 199] in knot-theoretic terms.

We now move to more general iterated torus knots. To proceed, recall the notion of type numbers following Definition 1.1.10. We call an iterated (horseshoe) torus knot of type $\{(p_i, q_i)\}$ (with $p_i < q_i$; $\forall i$) *resonant* if its type numbers q_i coincide with the periods q_i of the kneading invariants $\mathbf{w}_i$ in its defining sequence $\mathcal{W}$.

Theorem 4.3.12 (Holmes [88]) *Among the iterated horseshoe knots, each finite sequence $\{(p_i, q_i)\}_1^n$ of positive integers with p_i, q_i relatively prime and $p_i/q_i < \frac{1}{2}$ determines a unique pair of resonant iterated torus knots of type $\{(b_i, q_i)\}_1^n$ where $b_1 = p_1$ and*

$$b_{i+1} = q_{i+1} q_i b_i + (-1)^i p_{i+1}. \tag{4.10}$$

This result is essentially an iterated version of Theorem 4.2.9. It is proved by identifying the appropriate iterated horseshoe knots via their words and factored kneading invariants, and placing them correctly on the subtemplates of Theorem 4.3.8. The words are *-multiplied analogues of those for the simple torus knots of Theorem 4.2.9, and uniqueness follows by alternately maximizing and minimizing crossing numbers and appealing to Theorems 2.2.4 and 4.2.8. (Here, the embedded subtemplates are sufficiently twisted for one to apply Theorem 4.2.8 directly; no elaborate surgery as in Figure 4.6 is required.) The argument is lengthy and not particularly illuminating; for details and data on the associated kneading invariants and linking numbers see [88].

We close with a summary of orbit genealogies for the natural suspension of the horseshoe map. Generically, orbits appear as male-female pairs in saddle-node bifurcations, or as single female knots in period-doubling bifurcations. The female knots are "mothers," each of which forms the core of a subtemplate having the associated "father" knot as a boundary component. The mother is a companion (in the sense of Definition 1.1.10) to her infinitely many "children:" generalized cables which live on her subtemplate. Approximately half of these knots are female, and as such, proceed to form sub-subtemplates supporting infinitely many grandchildren, etc. Since each subtemplate is a twisted and (perhaps) knotted copy of the original, the bifurcation sequences on each subtemplate are miniature copies of the original but yield knots increasing in complexity. Not only are the individual orbits knotted and linked, but the sub-templates containing certain lineages of orbits are also twisted and linked about one another.

4.4 Homoclinic bifurcations

We now turn to some knot and link structures associated with global bifurcations involving homoclinic orbits to hyperbolic saddle points in three dimensional flows:

$$\dot{x} = f(x). \tag{4.11}$$

Suppose the saddle point lies at $x = 0$ ($f(x) = 0$) and let λ_i denote the eigenvalues of the linearization $Df(0)$. There are many possible cases to consider, for real and/or complex eigenvalues, and expanding or contracting flows, and we shall only give a brief sample of results. We start with the real, contracting case, summarising some results from [91], which the reader should consult for further detail.

4.4.1 Gluing and torus knots

Suppose that $Df(0)$ has three real eigenvalues with the single expanding eigenvalue $\lambda^u > 0$ weaker in magnitude than the two contracting eigenvalues: $-\lambda^{ss} > -\lambda^s > \lambda^u > 0$. We assume that both branches of the one-dimensional unstable manifold $W^u(0)$ lie in the two-dimensional stable manifold $W^s(0)$ and denote by Υ the set $W^u(0) \cup \{0\}$. This is a codmension two bifurcation, generically occurring at isolated points in parameter space for a two-parameter family of vector fields $f(x; \mu_1, \mu_2)$ (i.e., no symmetries are present). Letting $(\mu_1, \mu_2) = (0, 0)$ be such a point and varying (μ_1, μ_2), the degenerate case unfolds to the *gluing bifurcation*, in which up to two periodic orbits bifurcate from the double homoclinic loop Υ [63, 72].

Before stating the principal result, we must develop a little machinery. Denote the two loops of Υ: x_1 and x_2. The bifurcating periodic orbits may follow x_1 and/or x_2 many times before closing, giving a natural description as a word, much as in the symbolic description of templates. Those words which actually occur determine the unfolding. [63, 72] prove that any periodic orbit bifurcating from Υ must have a *rotation compatible* word.

Definition 4.4.1 An infinite (finite) word in two symbols is *rotation compatible* if it can be represented as the (finite periodic) itinerary of an orbit of a rigid rotation map $\rho_\theta : z \mapsto (z + \theta)$, $z \in S^1$, with the Markov partition $I(x_1) = (0, 1 - \theta]$, $I(x_2) = (1 - \theta, 1]$ for some $\theta \in [0, 1)$. The unique θ for such a word is its *rotation number*.

Remark 4.4.2 To compute the rotation number of a given finite rotation compatible word, take the number of x_2's and divide by the total length of the word: e.g., $x_1^2 x_2 x_1 x_2 \Rightarrow \theta = \frac{2}{5}$. The rotation compatible words are precisely the "evenly distributed" words of Theorem 4.2.9. Finally, we recall that two rational numbers $\frac{p}{q}$ and $\frac{p'}{q'}$ are *Farey neighbors* if $| pq' - qp' |= 1$.

Theorem 4.4.3 (Coullet et al. [63, 72]) *For every sufficiently C^1-small perturbation of $f(x; 0, 0)$ there are at most two periodic orbits in a small neighborhood N of Υ. Any such periodic orbits are attracting and have rotation compatible words, and, if there are two, their rotation numbers are Farey neighbors.*

The proof uses the eigenvalue condition, which implies that a small neighborhood of Υ is positively invariant and so contains an attractor, even after (small) perturbation. Defining cross sections near 0, one shows that the resulting return map is a (discontinuous) contraction. This, together with the fact that the attractor lies within the closure of the one-dimensional unstable manifold $W^u(0)$, of which there are two branches, implies that there are at most two stable periodic orbits for any given parameter pair (μ_1, μ_2). The admissible words are constructed via a reduced (one-dimensional) return map, which is effectively a discontinuous mapping of the circle. Note that there may be two, one, or *no* periodic orbits: both branches of the unstable manifold may limit on an "irrational" curve which winds repeatedly about, never closing.

Thus, unlike the expanding Lorenz flow, which is also related to a double homoclinic connection, gluing bifurcations create isolated periodic orbits characteristic of Morse-Smale flows (*cf.* Appendix A). The interest here is in describing how the rotation compatible periodic orbits succeed one another as the parameters (μ_1, μ_2) vary, and which knots and links they form. To determine the latter we will construct "templates" for the flows, relaxing the expansiveness demanded by the definitions of §2.2 to include contracting flows.

There are two distinct topological configurations, depending upon which sides of $W^s(0)$ the homoclinic orbits reenter: these are the *figure-of-eight* and the *butterfly*, shown in Figure 4.12. For both systems, we assume the existence of a strong stable foliation (reported in [72] to be a generic condition in these cases) and collapse out as in the proof of Lemma 2.2.7, leaving a (contracting) template. Alternatively, these branched manifolds may be viewed as embedded suspensions of one-dimensional noninvertible return maps.

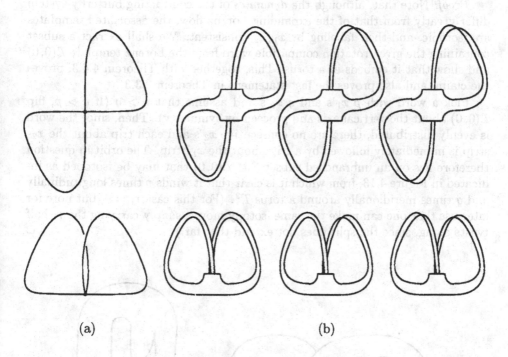

(a)   (b)

Figure 4.12: (a) The figure-of-eight and butterfly configurations, and (b) associated templates.

Embedding these templates in $\mathbb{R}^3$, we must incorporate the "twist" of the flow around the homoclinic connections, which leads to twisting of the template strips. Temporarily ignoring full (even) twisting of each strip and excluding non-trivially knotted embeddings, there are three intrinsic cases to consider: *untwisted*: $\tau_1 = \tau_2 = 0$; *singly-twisted*: $\tau_1 = 0$, $\tau_2 = 1$; and *doubly-twisted*: $\tau_1 = \tau_2 = 1$, also illustrated in Figure 4.12. Below we give results only for the butterfly case: the figure-of-eight, whose template is unbranched, is somewhat simpler. For details see [91].

Case (1) untwisted: $\tau_1 = 0$, $\tau_2 = 0$

Using the theory of circle maps (one views the Poincaré map as a monotone injective map of the circle with a single discontinuity), in [72, 65], it is proved that this system has at most one periodic orbit. As an addendum to this, we have:

Proposition 4.4.4 *Any periodic orbit appearing in the unfolding of an untwisted butterfly is a torus knot. If the rotation number of the word is* $\theta = \frac{p}{q+p}$, *then the corresponding knot type is* (p, q).

Proof: Note that, although the *dynamics* of the contracting butterfly system differ greatly from that of the expanding Lorenz flow, the associated templates are isotopic, and their labeling by x_1, x_2 consistent. We shall extract a subset containing the given rotation compatible word from the Lorenz template $\mathcal{L}(0,0)$ and show that it embeds in a torus. This, together with Theorem 4.4.3, proves the claim, and also proves the last statement in Theorem 2.3.3.

Pick a word with p x_1's and q x_2's and assume that $p > q$ (If $q > p$, flip $\mathcal{L}(0,0)$ about the vertical axis and proceed by symmetry). Then, since the word is evenly distributed, there are no consecutive x_2's and each trip about the x_2-strip is immediately followed by a trip about the x_1-strip. The orbit in question therefore lies on an unbranched subset $S \subset \mathcal{L}(0,0)$ that may be isotoped as indicated in Figure 4.13, from which it is clear that it winds p times longitudinally and q times meridionally around a torus T^2. (For this case $\tau_1 = 0$, but note for later use that one can make the same isotopy moves, simply carrying the τ_1 half twists along, since the split does not extend that far.) $\square$

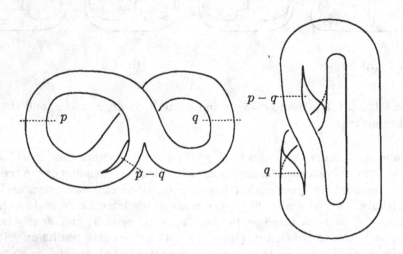

Figure 4.13: The subset $S \subset \mathcal{L}(0,0)$ fits on a torus T^2. The labels refer to the number of strands on each strip.

Example 4.4.5 The words $x_1^2 x_2 x_1 x_2$ and $x_1 x_2^2 x_1 x_2^3$ correspond to $(2,3)$ and $(5,7)$ torus knots respectively. Note that the mapping from words to torus knots differs from that on the horseshoe template $\mathcal{H} = \mathcal{L}(0,1)$, described in the proof of Theorem 4.2.9.

Observe that this result merely proves that *if* an orbit with the given word exists, then it is a torus knot of the type specified. To find such orbits, one has to tune the parameters (μ_1, μ_2) appropriately, as specified in the bifurcation diagrams of [72, 65] and summarised in [91]. Between each pair of (disjoint) open sets (μ_1, μ_2) giving rise to torus knots of Farey neighbor types (p, q), (p', q'), there is a set having knots of type $(p+p', q+q')$: the *Farey mediant*. In this way, passing across the parameter plane, one exhausts all torus knots. Intuitively, we are moving the thin incoming strips along the branch line of the contracting template to match up, one by one, the "ends" of the torus knots which all coexist on the expanding Lorenz template $\mathcal{L}(0, 0)$.

We briefly consider the impact of introducing τ_1 (even) positive half-twists along the x_1 branch. The proof of Proposition 4.4.4 may be modified to cope with this case, as already indicated. Even if τ_1 is non-zero we may perform the same moves without interference from the additional half-twists. Then, since τ_1 is even and there are $\frac{1}{2}\tau_1$ full twists, we obtain a $(p, q + \frac{1}{2}p\tau_1)$ torus knot (to check this, refer to the positive braid genus formula of Equation (1.4)). A similar argument for $\tau_1 = 0$ and τ_2 even yields a $(p + \frac{1}{2}q\tau_2, q)$ torus knot.

If both τ_1 and τ_2 are simultaneously non-zero and even, the resulting subset S can still be presented as a positive braid on p strands, but it is no longer a torus knot, for there is additional twisting on the strip carrying q strands. Indeed, it does not appear to belong to any well-known knot family. A picture and genus formulae for this case appear in [91].

Case (2) singly- and doubly-twisted: $\tau_1 = 0, 1; \ \tau_2 = 1$

In these cases one can use contraction and orientation-reversal properties of the one-dimensional return map induced by the semiflow, along with template surgery analogous to that of Figure 4.13, to prove the following rather restrictive result:

Proposition 4.4.6 ([91]) *If the x_2-branch of the butterfly template has a half-twist (case (2)) then all periodic orbits appearing on it must have words x_1 or $x_1^k x_2$ ($k \geq 0$). The same holds reversing x_1 and x_2. If both branches have half twists (case (3)), then only x_1, x_2, and $x_1 x_2$ may appear. Any periodic orbit appearing in the unfolding of either case is an unknot.*

Remark 4.4.7 The significance in the knotting and linking of orbits implicated in gluing bifurcations lies not so much in extracting bifurcation invariants (for these bifurcations are fairly well-understood), but in displaying the general principle that simple dynamics are coupled with the existence of simple knots and links. The fact that only torus knots can occur in a butterfly-gluing bifurcation (in which the flows are all zero-entropy) is in stark contrast to the analogous positive entropy Lorenz flow, in which an infinite array of knot types coexist: *cf.* Theorems 3.1.15 and A.1.13.

The next example of global bifurcations exhibits an opposite extreme of topological complexity.

4.4.2 Silnikov connections and universal templates

We now return to the example presented in §2.3.3: a radically different type of global bifurcation, originally studied by Shil'nikov [160, 161] (*cf.* [179] and the textbooks [76, 188, 189], which also contain these and related results). The material below is adapted from [71]. Recall the definition of a *Shil'nikov connection*, Definition 2.3.8, and the associated Theorem 2.3.9: that a countable collection of suspended horseshoes lives in a tubular neighborhood of a Shil'nikov connection.

Sketch of proof of Theorem 2.3.9: We construct Poincaré sections transversal to Γ near the fixed point p and linearize the flow near p and along Γ to obtain approximate return maps. The horseshoes are constructed by flowing pairs of boxes near p and then along Γ. The fixed point has a one-dimensional unstable manifold $W^u(p)$ and a two-dimensional stable manifold $W^s(p)$, along which $\Gamma = W^s(p) \cap W^u(p)$ spirals into p. (Although we consider only the case in which $W^u(p)$ is one-dimensional, our results apply equally well to $W^u(p)$ two-dimensional and $W^s(p)$ one-dimensional, since this amounts to a reversal of time which leaves periodic orbits invariant.)

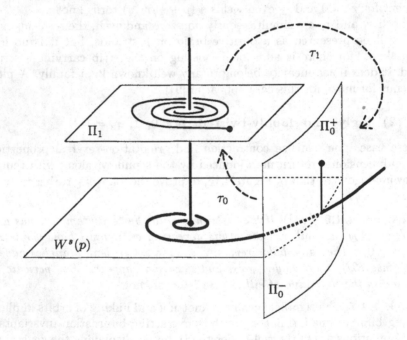

Figure 4.14: Cross sections and maps near the fixed point p.

We construct Poincaré sections Π_0 and Π_1 transverse to Γ and sufficiently close to p that linear analysis provides a good estimate of the return map. The surface Π_0 is bisected by $W^s(p)$ into upper (Π_0^+) and lower (Π_0^-) halves. We use a cylindrical coordinate system having origin at p and with Π_0 at constant r and Π_1 at constant $z = \epsilon \ll 1$ (this is the convention of [76] — one may just as well choose Π_0 at constant θ [74, 189]): see Figure 4.14. The return map factors into the "local" map $\tau_0 : \Pi_0^+ \to \Pi_1$, which is concentrated near p, and the "global" map $\tau_1 : \Pi_1 \to \Pi_0$, which follows orbits along near Γ. Hypotheses (1) and (2) permit us to construct approximations to these maps.

Taking Π_0 and Π_1 close enough to p, the flow linearised at p,

$$
\begin{aligned}
r(t) &= r_0 e^{-\lambda^s t} \\
\theta(t) &= \theta_0 + \omega t \\
z(t) &= z_0 e^{\lambda^u t},
\end{aligned}
\tag{4.12}
$$

provides a good approximation of τ_0. Solving $z(T) = \epsilon$ for T, we obtain the transit time for orbits leaving Π_0 to reach Π_1:

$$
T(z) = \frac{1}{\lambda^u} \log \frac{\epsilon}{z}.
\tag{4.13}
$$

This yields an expression for the local return map τ_0:

$$
\tau_0 : (r_0, \theta, z) \mapsto \left(r_0 \left(\frac{\epsilon}{z} \right)^{\lambda^s / \lambda^u}, \theta + \frac{\omega}{\lambda^u} \log \left(\frac{\epsilon}{z} \right), \epsilon \right).
\tag{4,14}
$$

Restricting to a sufficiently small neighborhood of $\Gamma \cap \Pi_1$, one can assume that the global return map τ_1 is affine. This yields an analytical approximation to the Poincaré map given by composition of (4.14) with an affine map. Such composed maps have been analyzed repeatedly [160, 161, 74, 66].

The action of τ_0 on a segment of constant θ is to stretch it and wrap it around $\Gamma \cap \Pi_1$ in a logarithmic spiral. Since $z = 0$ is on $W^s(p)$, the image of $\tau_0(r, z)$ as $z \to 0$ approaches $\Gamma \cap \Pi_1$. This image is then mapped affinely back to Π_0, with $\tau_1(\Gamma \cap \Pi_1) = \Gamma \cap \Pi_0$: see Figure 4.14.

One now examines the action of $\tau_1 \tau_0$ on rectangular strips:

$$
B_i = \{ (\theta, z) \subset \Pi_0^+ \; : \; a_i \le z \le b_i \},
\tag{4.15}
$$

where the sequences $\{a_i\}$ and $\{b_i\}$ satisfy $a_i < b_i < a_{i-1}$ and $\lim_{i \to \infty} a_i = 0$. For appropriate choice of numbers $\{a_i, b_i\}$, it can be shown [76, 188, 189] that the image of each adjacent pair $\{B_i \cup B_{i+1}\}$ under $\tau_1 \tau_0$ intersects $B_i \cup B_{i+1}$ to form a hyperbolic horseshoe (see e.g. Theorem 4.8.4 of [189]): see Figure 2.11(b). These pairs are the horseshoes of Theorem 2.3.9. □

We now develop a geometric treatment based on the analysis sketched above (*cf.* [5]), which will allow us to extract the desired templates and prove that the flow in a neighborhood of a double Shil'nikov connection contains representatives of all knots and links.

Single Shil'nikov templates

The horseshoes of Theorem 2.3.9 are hyperbolic, so we may collapse along the stable foliations and, carefully following the embedding, construct the embedded template. We proceed in two steps, according to the two components of the return map $\tau_1\tau_0$.

First, the action of the global map τ_1 is affine and takes the image under τ_0 of the "horizontal" $B_i \subset \Pi_0^+$ to a "vertical" strip in Π_0. Collapsing in the contracting direction of the map $\tau_0\tau_1$, each box $B_i \subset \Pi_0^+$ becomes a vertical interval $\{a_i \leq z \leq b_i\}$ at a fixed r. Thus, the collapsed B_i and B_{i+1} boxes are disjoint within Π_0^+. Their images, however, are vertical lines which cover Π_0; hence, the two strips are joined at a branch line.

Since τ_1 is affine, there is no additional folding. Therefore, instead of collapsing the stable direction out to obtain a branch line in Π_0^+, we can propagate the branch line back via τ_1^{-1} to depict the joining of these strips within Π_1, as in Figure 2.12(b). The impact of τ_1 on the topology of the suspension is encoded in the *twist* of Γ between Π_1 and Π_0^+ (*cf.* Remark 1.2.18). For N a small tubular neighborhood of Γ excluding a small neighborhood of p, $W^s(p) \cap N$ is a two-dimensional strip which may twist any number of times about Γ. Since $\tau_1\tau_0(B_i)$ transversally intersects $W^s(p)$, the template inherits this same twist: see again Figure 2.12(b).

The action of the local map, τ_0, is to stretch B_i out along what was the z-direction in Π_0^+ and compress B_i along what was the θ-direction. The image of $\tau_0(\Pi_1^+)$ is a thin spiral (imagine thickening that in Figure 4.14). The image of any consecutive pair B_i, B_{i+1} lies within a folded strip: a horseshoe. As the box $B_i \subset \Pi_1^+$ flows through a neighborhood of p to reach Π_1, it is wrapped around Γ an integer number of half-turns, B_{i+1} being wrapped with one more half-turn than B_i. Indeed, the winding which occurs near p is revealed by Eqn. (4.14). As detailed in [188, 189], the boxes B_i can be chosen such that

$$a_i = \epsilon e^{-\pi i \lambda^u / \omega}. \tag{4.16}$$

Hence,

$$\Delta\theta \;\;\equiv\;\; \omega(T(a_{i+1}) - T(a_i)) \tag{4.17}$$

$$=\;\; \frac{\omega}{\lambda^u}\left(\log\frac{\epsilon}{a_{i+1}} - \log\frac{\epsilon}{a_i}\right) = \pi,$$

and the action of the flow of B_{i+1} from Π_0^+ to Π_1 is to wind about Γ in the θ direction by an additional π, compared to B_i. This is shown in Figure 4.15.

Remark 4.4.8 The strips drawn in Fig. 4.15 are shown with minimal twisting; however, there is no guarantee that the "topmost" B_i, which suffer the least twist, satisfy the hyperbolicity conditions necessary for Theorem 2.3.9. We only know that for i (and hence, twist) sufficiently large, pairs of boxes $B_i \cup B_{i+1}$ can be chosen so that their images form hyperbolic horseshoes.

We may now classify the types of horseshoe templates which appear near Γ. For i some fixed integer, consider the template formed by collapsing the contracting directions of the flow of the boxes B_i and B_{i+1}. In a neighborhood of p, the strip corresponding to B_i (resp. B_{i+1}) winds about Γ with i (resp.

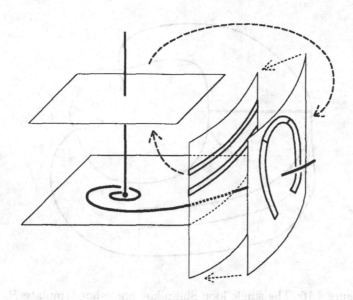

Figure 4.15: A "simple" Shil'nikov horseshoe.

$i + 1$) half-twists. The strips join at Π_1 in a single strip which follows Γ back to Π_0, undergoing a further M half-twists, for some fixed (but unknown) M.

If we assume that the homoclinic connection is unknotted, the template thus obtained depends only on the depth of the horseshoe, i, and the fixed global twisting, M. Up to homeomorphism, there are two types, depending upon the parity of $\alpha \equiv i + M$. The template $\mathcal{H}_\alpha$ is shown in Figure 4.16: for α even, this is homeomorphic (though not isotopic!) to the standard horseshoe template $\mathcal{H}$ (*cf.* Figure 2.9), and for α odd, this is homeomorphic to the "twisted" horseshoe template $\tilde{\mathcal{H}}$ of Figure 4.10. For any α, $\mathcal{H}_\alpha$ is isotopic to $\mathcal{H}$ with α additional half-twists inserted after the branch line.

For a given flow, the global twisting M and the minimum depth i of its horseshoes are effectively uncomputable; hence, one cannot rigorously conclude the existence of any *particular* $\mathcal{H}_\alpha$ for a fixed system, only for α greater than some (unknown) lower bound. We will now bypass this problem by considering a double connection which induces equal positive and negative twisting and cancelling the two unknown twists.

Double Shil'nikov templates

Definition 4.4.9 A function $f : \mathbb{R}^n \to \mathbb{R}^n$ is *equivariant* with respect to a function $\Psi : \mathbb{R}^n \to \mathbb{R}^n$ if $\Psi f(\mathbf{x}) = f(\Psi(\mathbf{x}))$ for all $\mathbf{x} \in \mathbb{R}^n$.

We shall consider Shil'nikov connections in which the vector field of the differential equation $\dot{\mathbf{x}} = f(\mathbf{x})$ is equivariant under a symmetry of one of the

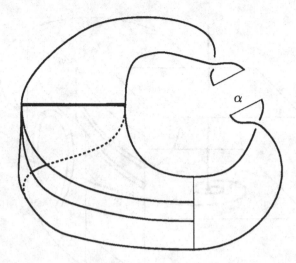

Figure 4.16: The single loop Shil'nikov horseshoe template $\mathcal{H}_\alpha$.

following forms:

$$\Psi : (x, y, z) \mapsto (-x, -y, -z)$$
$$\Psi : (x, y, z) \mapsto (-x, -y, z)$$
(4.18)

Such symmetries are quite common: the Lorenz system exhibits the second type [114, 76]. If the system additionally has a fixed point, p, satisfying the conditions of Theorem 2.3.9, the flow will appear as one of the three cases shown in Figure 4.17, displaying either a pair of homoclinic spirals at $p = \Psi(p)$, or a spiral *heteroclinic cycle* connecting p and $\Psi(p) \neq p$. Naturally, an analogue to Theorem 2.3.9 holds in this case, with the added ingredient of "coupled horseshoes" [75, 86, 16].

We now extend the arguments given above for the single loop case to the double loop homoclinic orbit of Figure 4.17 [left], having the first symmetry of equation (4.18), so that the loop Γ has a partner $\Gamma' = \Psi(\Gamma)$. (The other heteroclinic cases can be dealt with similarly: see [71] for details.) As in the single loop case of Figure 4.14, we define Poincaré sections Π_0 and Π_1, but now along with their images under Ψ: Π_0' and Π_1'. Note that Π_1 is above the saddle and Π_1' below, and Π_0 and Π_0' on opposite sides. Using the same linear and affine approximations as before, we derive two local and two global return maps τ_0 and τ_0' and τ_1 and τ_1', but in this case we define strips $B_i \subset \Pi_0$ and $B_i' \subset \Pi_0'$, so that $\tau_0(B_i) \subset \Pi_0$, $\tau_0'(B_i') \subset \Pi_0'$, $\tau_1(\Pi_1) \subset \Pi_0'$ and $\tau_1'(\Pi_1') \subset \Pi_0$. Thus we restrict our attention to orbits which make double traverses of a neighborhood of $\Gamma \cup \Gamma'$, tracking the two loops in regular succession.

Following the construction for the single loop case, we produce the template of Figure 4.18, in which the strip leaving the upper branch line in Π_1 connects to Π_0', and that leaving the lower branch line in Π_1' connects to Π_0. The resulting

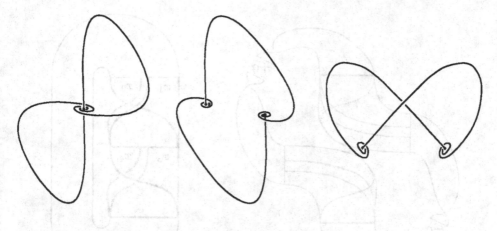

Figure 4.17: Three symmetric homoclinic configurations.

template has two branch lines and contains a copy of the single loop template $\mathcal{H}_\alpha$ of Figure 4.16 followed by its image under Ψ. Since Ψ reverses orientation $(\det(D\Psi = -1))$, the sense of twist in these two components is opposite; indeed, whatever the depth i, we may collect all the "extra" twisting of the upper component as a group of $\alpha = i + M$ positive half twists and that of the lower as α negative half twists. These twists may clearly be cancelled exactly, leaving a pair of "simple" horseshoe templates, one positive and one negative, as shown in Figure 4.18. We call the resulting template $\mathcal{Z}$.

Remark 4.4.10 We assume that the homoclinic/heteroclinic connections involved in the double Shil'nikov connection are unknotted. Otherwise, the template $\mathcal{Z}$ might be nontrivially knotted, obstructing our final step below.

ODEs which generate all knots and links

The template $\mathcal{Z}$, which appears near the double Shil'nikov loop, shares the richenss of the templates of §3.2:

Lemma 4.4.11 *The template $\mathcal{Z}$ is universal: it contains an isotopic copy of every knot and link.*

Proof: The symbolic inflation $\Im$ given by

$$\Im : \mathcal{V} \hookrightarrow \mathcal{Z} \qquad \begin{cases} x_1 \mapsto x_2 x_4 \\ x_2 \mapsto x_1 \\ x_3 \mapsto x_4 x_2 \\ x_4 \mapsto x_3 \end{cases} , \qquad (4.19)$$

defines a map from $\mathcal{V}$ into $\mathcal{Z}$. The astute reader will note that the images of the periodic orbits $(x_1)^\infty$ and $(x_3)^\infty \in \mathcal{V}$ map to $(x_2 x_4)^\infty = (x_4 x_2)^\infty$ in $\mathcal{Z}$: the

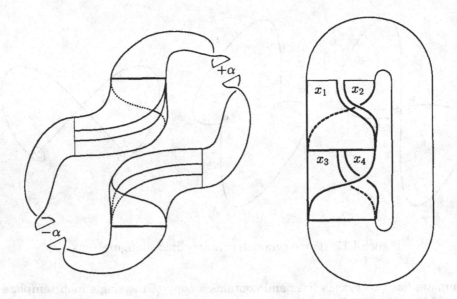

Figure 4.18: The double loop Shil'nikov horseshoe template $\mathcal{Z}$, before (left) and after (right) cancelling the opposite twists.

same orbit. While this precludes Equation (4.19) from satisfying the definition of an inflation (the image is not a proper subtemplate), we may nevertheless disregard this anomaly by performing a DA-splitting of $\mathcal{Z}$ along $(x_2x_4)^{\infty}$ and proceeding as usual. The orbit $(x_2x_4)^{\infty}$ is an unknot and there are many more unknots in the template. Figure 4.19 shows that the subtemplate defined by $\mathcal{I}$ is isotopic to $\mathcal{V}$. $\square$

As a corollary, we obtain the following remarkable:

Theorem 4.4.12 *Sufficient conditions for a third-order ODE to contain periodic orbits representing all knot and link types are that the vector field is sufficiently C^1-close to a vector field satisfying the following four conditions:*

1. *There exists a fixed point p for the vector field, and the linearization $Df|_p$ at p has eigenvalues $\{-\lambda^s \pm \omega i, \lambda^u\}$, with*

$$\lambda^u > \lambda^s > 0 \qquad \omega \neq 0. \tag{4.20}$$

2. *The flow ϕ_t is equivariant under one of the following symmetries:*

$$\begin{aligned} \Psi &: (x,y,z) \mapsto (-x,-y,-z) \\ \Psi &: (x,y,z) \mapsto (-x,-y,z) \end{aligned}. \tag{4.21}$$

3. *There exists an orbit $\Gamma(t)$ with $\lim_{t\to-\infty} \Gamma(t) = p$ and $\lim_{t\to\infty} \Gamma(t) = \Psi(p)$.*

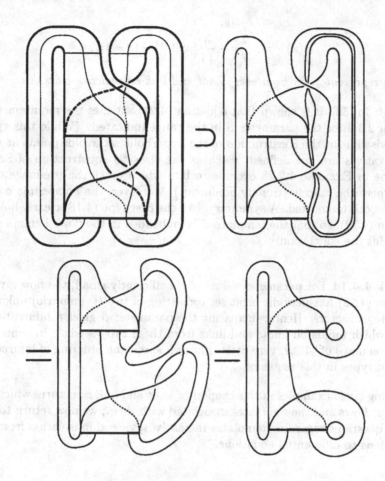

Figure 4.19: $\mathcal{V}$ is a subtemplate of $\mathcal{Z}$.

4. The homoclinic/heteroclinic loop(s) is(are) unknotted.

The constructions preceeding the proof of Theorem 4.4.12 above actually show that, as one approaches the degenerate double loop, one can pick successively smaller tubular neighborhoods of the double loop which contain infinitely many copies of representatives of every knot and link equivalence class.

Thanks to the work of Chua et al. [38], we can even display an explicit example of a three-dimensional system which contains a universal template:

Corollary 4.4.13 *There exists an open set of parameters $\beta \in [6.5, 10.5]$ for which the set of periodic solutions to the differential equation*

$$
\begin{aligned}
\dot{x} &= 7[y - \phi(x)], \\
\dot{y} &= x - y + z,
\end{aligned}
$$

$$\dot{z} \;=\; -\beta y, \tag{4.22}$$

$$\phi(x) \;=\; \frac{2}{7}x - \frac{3}{14}\left[\,|x+1| - |x-1|\,\right],$$

contains representatives from every knot and link equivalence class.

Proof: In [38], it is shown that equation (4.22) satisfies the requirements of Theorem 2.3.9 for the parameter β in the range indicated. (While this system is piecewise linear, the construction of the hyperbolic set avoids points at which the derivatives are not defined, much as the classical construction of Smale's horseshoe in Example 1.2.28 excludes orbits which enter the preimage of the bend, where the map is strongly nonlinear.) Moreover, the homoclinic connections are both unknotted. A symmetry Ψ of the first type (4.18) clearly holds for Equation (4.22), so that the template $\mathcal{Z}$ is embedded in the flow. Lemma 4.4.11 then yields the conclusion. $\qquad\qquad\qquad\qquad\qquad\qquad\qquad\qquad\qquad\quad\Box$

Remark 4.4.14 For parameter values of β sufficiently small, the flow given by Equation (4.22) has periodic orbit set consisting of two (symmetric) unknotted separable attractors. Hence, increasing the parameter β gives a bifurcation sequence which builds all knots and links from these two "seeds." In contrast to the Hénon maps of §4.2.2, very little is known about the ordering of bifurcations and knot types in this sequence.

Having given examples in this chapter of knot and link structures which arise in specific flows and the templates associated with them, we now return to more general questions regarding templates themselves, viewed in isolation from their connections to differential equations.

Chapter 5: Invariants

Recall the fundamental problem in knot theory: when are two knots (links) equivalent? An analogous problem presents itself: when are two templates equivalent? We must first, however, carefully state what equivalence we want, since we are chiefly interested in the knots and links that inhabit a template, as opposed to the branched manifold itself. With this is mind, we proceed with a suitable definition of equivalence.

Recall that many orbits in a template's semiflow exit the template. Periodic orbits of course remain on the template forever, but so do asymptotically periodic and certain other orbits. Those points whose forward trajectories never exit the template comprise the chain-recurrent set of the template (*cf.* Definition 1.2.11 and the orbits which never leave the Smale horseshoe map.)

Definition 5.0.1 Two embedded templates in S^3 are *equivalent* if they are connected by a finite sequence of the following template "moves:"

1. Ambient isotopy on the template;

2. The *split* move; and

3. The *slide* move.

The split and slide moves are illustrated in Figure 5.1.

Remark 5.0.2 The reader might feel the slide move is just an isotopy. But, when the branch lines momentarily coincide, the object obtained is not technically a template according to Definition 2.2.1.

Remark 5.0.3 All three of the above moves induce an isotopy on the chain-recurrent set of a template.

The standard invariants of topology (*e.g.*, the fundamental group) are altered by the split move. Hence, we must search for other means to construct invariants of templates. We give two brief examples of template invariants which are topological in nature.

Perhaps the simplest invariant is *orientability*. By *orientation* we mean a coordinate system that can be translated about by the flow. The horseshoe template $\mathcal{H}$ contains a smooth Möbius strip of flow lines, and hence is *nonorientable as a template*. The Lorenz template is orientable in this sense. No finite sequence of template moves can take an orientable template to a nonorientable template.

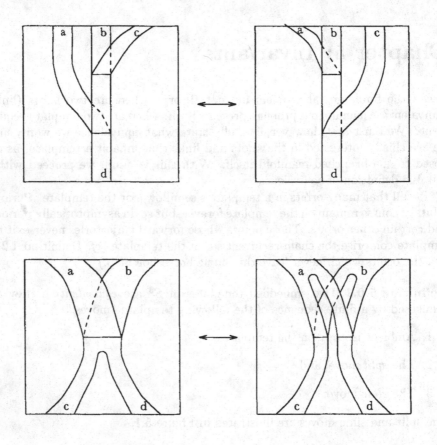

Figure 5.1: Template moves: slide (above) and split (below)

The link of closed orbits in the boundary of a template (perhaps empty) is not changed by either template move and is thus an invariant. Even the framing of the *boundary link* is invariant: the twisting of the unit tangent bundle restricted to the boundary link is unchanged by template moves. Other loops in the boundary of a template can be used to produce invariants. Consider loops with one cusp (see Figure 5.2). The split move can only create or destroy loops with two cusps. However, we need to be careful in how we count loops with one cusp; we can use the cusp only once. Otherwise the split could affect the counting of one cusp loops. In fact for every $n \neq 2$ the number of boundary loops with n cusps is an invariant. Of course, all this requires that the charts be attached smoothly and that the exit sets of the split charts be smooth. This can always be done. We record these observations below.

Lemma 5.0.4 *Given $\mathcal{T} \subset S^3$ an embedded template, the set of closed orbits which lie within the boundary of $\mathcal{T}$, considered as a framed link, is an invariant of $\mathcal{T}$. Furthermore, if we consider $\partial \mathcal{T}$ as a smooth graph, then loops which do* *not have exactly two cusp points are invariant.*

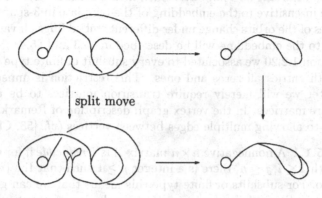

Figure 5.2: Counting boundary loops

Example 5.0.5 The Lorenz template has two unknotted unlinked orbits in its boundary. The horseshoe template has one closed orbit and one loop with a single cusp; these loops are also unknotted and unlinked.

Corollary 5.0.6 *A complete template invariant yields a complete knot invariant.*

Proof: Given any knot K, let $\mathcal{T}_K$ denote the embedded template obtained from the horseshoe template by re-embedding the x_1 strip so that the orbit x_1^∞ has knot type K with zero twist. Then, since the boundary link of $\mathcal{T}$ is precisely the knot K, the ability to distinguish any two such templates implies the ability to distinguish the boundary knots.   □

In the next section, we begin with an invariant derived solely from dynamical data (*i.e.*, the embedding of the template is not considered). In §5.2, we extend this invariant to one which accounts for orientations of the strips in a template. Then, in §5.3, we turn to the ζ-function of a flow as a means of counting twists of embedded orbits, thereby constructing a dynamical invariant sensitive to embedding. In §5.4 we discuss another type of ζ-function that encodes linking information in Lorenz templates.

5.1 Classifying suspended subshifts

The underlying dynamics on a template are the suspended subshifts of finite type, as discussed in §2.2. Two suspensions of subshifts of finite type are *topologically equivalent* if there is a homeomorphism between them that takes orbits to orbits and preserves the flow direction. Our goal in this section is to describe a classification theorem for suspensions of subshifts of finite type with respect to flow equivalence. Any invariant of suspensions of subshifts of finite type is automatically an invariant for templates. Such invariants are abstract in the sense that they are insensitive to the embedding of the template in 3-space. Of course the knot types of the orbits change under different embeddings. Invariants which are sensitive to the embedding will be described in §5.3 and §5.4.

In Definition 1.2.20 we associated to every subshift of finite type a transition matrix A with entries all zeros and ones. This restriction is unnecessary and in this chapter we will merely require transition matrices to be nonnegative integral square matrices. In the vertex graph description of Remark 1.2.22, this is equivalent to allowing multiple edges between vertices (cf. [53, Chapter 3]).

Definition 5.1.1 A nonnegative $n \times n$ matrix A is *irreducible* if for each integer pair (i, j) with $1 \leq i, j \leq n$, there is a integer $p \geq 1$ such that the (i, j) entry in A^p is nonzero. For subshifts of finite type this means that we can get from any given Markov partition element to any other (or the same) partition element by iterating the shift map σ.

Irreducible transition matrices correspond to subshifts of finite type with a dense orbit (cf. Corollary 3.1.17); that is, there is a single basic set.

Definition 5.1.2 Two nonnegative square integer matrices, A and B are *strong shift equivalent* $A \overset{s}{\sim} B$, if there exist nonnegative square integer matrices $A = A_1, \ldots, A_{k+1} = B$ and nonnegative integer (not necessarily square) matrices $R_1, S_1, \ldots, R_k, S_k$ such that $A_i = R_i S_i$ and $A_{i+1} = S_i R_i$ for $i = 1, \ldots, k$.

This "move" corresponds to making certain changes in the choice of the Markov partition. Roughly speaking we can relabel partition elements, refine them (*i.e.*, choose smaller disks) or combine them (*i.e.*, choose bigger disks). The next theorem asserts that this suffices to generate conjugacy.

Theorem 5.1.3 (Williams [191]) *Suppose A and B are nonnegative square integer matrices and σ_A and σ_B are the corresponding subshifts of finite type. Then σ_A is topologically conjugate to σ_B if and only if A is strong shift equivalent to B.*

A concise proof of Theorem 5.1.3 can be found in [53, Appendix A].

Remark 5.1.4 Any nonnegative square integer matrix is strong shift equivalent to a square matrix whose entries are just zeros and ones.

Example 5.1.5 Let $A = \begin{bmatrix} 1 & 1 \\ 1 & 0 \end{bmatrix}$, and $B = \begin{bmatrix} 1 & 1 & 0 \\ 0 & 0 & 1 \\ 1 & 1 & 0 \end{bmatrix}$.

Then using $R = \begin{bmatrix} 1 & 1 & 0 \\ 0 & 0 & 1 \end{bmatrix}$, and $S = \begin{bmatrix} 1 & 0 \\ 0 & 1 \\ 1 & 0 \end{bmatrix}$, we get $A = RS$ and $B = SR$. In this example the sequence length, sometimes called the *lag*, was just one — such luck is rare.

Exercise 5.1.6 Show that $[2] \overset{s}{\sim} \begin{bmatrix} 1 & 1 \\ 1 & 1 \end{bmatrix}$.

Exercise 5.1.7 Prove that any relabeling of the elements of a Markov partition can be realized by strong shift equivalence.

Two irreducible nonnegative square integral matrices are *flow equivalent* if the suspensions of the corresponding subshifts of finite type are topologically equivalent. The suspension of a subshift of finite type corresponding to a permutation matrix is a finite collection of closed orbits. Irreducible permutation matrices are thus said to form the *trivial flow equivalence class*. In order to characterize the flow equivalence classes of irreducible nonnegative square matrices we need an additional "move" know as *expansion equivalence*. The idea is that we can change a Markov partition by adding a new partition element "parallel" to an current one. That is the new partition element is a forward (or backwards) translation via the flow of a current partition element.

Definition 5.1.8 Two square matrices A and B are *expansion equivalent*, $A \overset{e}{\sim} B$, if

$$A = \begin{bmatrix} a_{11} & \cdots & a_{1n} \\ \vdots & & \vdots \\ a_{n1} & \cdots & a_{nn} \end{bmatrix} \text{ and } B = \begin{bmatrix} 0 & a_{11} & \cdots & a_{1n} \\ 1 & 0 & \cdots & 0 \\ 0 & a_{21} & \cdots & a_{2n} \\ \vdots & \vdots & & \vdots \\ 0 & a_{n1} & \cdots & a_{nn} \end{bmatrix},$$

or *vice versa*.

Here $A \overset{e}{\sim} B$ represents expansion along the first partition element. But, since renumbering the partition elements can be realized by strong shift equivalence, this is the only expansion we need consider.

Parry and D. Sullivan showed that these two moves — strong shift equivalence and expansion equivalence — generate flow equivalence [141].

Theorem 5.1.9 (Parry and Sullivan [141]) *Two nonnegative square integer matrices A and B are flow equivalent if and only if there exist a finite sequence of square nonnegative matrices $A = A_0, A_1, \ldots, A_r = B$ with $A_i \overset{s}{\sim} A_{i+1}$ or $A_i \overset{e}{\sim} A_{i+1}$ for $i = 0, \ldots, r - 1$.*

As a corollary, we obtain our first dynamical invariant.

Corollary 5.1.10 *If A and B are flow equivalent then $\det(I - A) = \det(I - B)$.*

Proof: The proof is an exercise, though beware of sign errors.   □

Bowen and Franks [27] developed another invariant of suspensions of subshifts of finite type, working at least initially from a different point of view. Using an $n \times n$ transition matrix A they consider the group

$$G_{I-A} = \mathbb{Z}^n / (I - A) \mathbb{Z}^n.$$

Theorem 5.1.11 (Bowen and Franks [27]) *If A and B are flow equivalent then $G_{I-A} \cong G_{I-B}$.*

Outline of proof: Let A be an $n \times n$ integer matrix. Consider the action of A on the n-torus T^n. The fixed points of A form a subgroup of T^n under vector addition (mod 1). The fixed point subgroup is also given by the kernel of the map $(I - A) : T^n \to T^n$. By a standard duality theorem the kernel is isomorphic to the co-kernel of the map $(I - A) : \mathbb{Z}^n \to \mathbb{Z}^n$, which is just G_{I-A}.

Under strong shift equivalence the fixed point set of A is unchanged. For the expansion move one shows that it is equivalent to taking a direct sum with a trivial group and so does not effect the isomorphism class.   □

We can now state the classification theorem:

Theorem 5.1.12 (Franks [55]) *Suppose that A and B are nonnegative irreducible integer matrices, neither of which is in the trivial flow equivalence class. The matrices A and B are flow equivalent if and only if*

$$\det(I_n - A) = \det(I_m - B)$$

and

$$\frac{\mathbb{Z}^n}{(I_n - A)\mathbb{Z}^n} \cong \frac{\mathbb{Z}^m}{(I_m - B)\mathbb{Z}^m},$$

where n and m are the sizes of A and B respectively, I_n and I_m are identity matrices, and $\cong$ denotes group isomorphism.

Remark 5.1.13 Theorem 5.1.12 does not hold if the trivial flow equivalence class is not excluded.

Theorem 5.1.12 does not have a very good resolution for distinguishing templates. Consider the Lorenz and Horseshoe templates ($\mathcal{L}(0,0)$ and $\mathcal{H}$ from §2.3). These each have the matrix

$$\begin{bmatrix} 1 & 1 \\ 1 & 1 \end{bmatrix}$$

as a transition matrix, yet surely they are not equivalent, since $\mathcal{H}$ is not orientable while $\mathcal{L}(0,0)$ is: no finite sequence of template moves transforms an orientable template into a nonorientable template.

5.1.1   Finitely generated Abelian groups

It is worth noting that although strong shift equivalence is not generally computable, the invariants of suspensions of subshifts of finite type are readily computed. To see this we digress briefly into the theory of Abelian groups. Any square integer matrix A yields an Abelian group

$$G_A = \mathbb{Z}^n / A\mathbb{Z}^n,$$

where n is the size of A. However, different matrices can give rise to isomorphic groups. If matrix B can be obtained form matrix A by a finite sequence of operations (to be listed shortly) then $G_A \cong G_B$. The matrices A and B do not need to be the same size. Furthermore, each isomorphism class of matrices has a canonical representative which can be computed from any other matrix in its class by a finite algorithm; thus, the converse holds as well. The allowed operations are:

- switching two rows,

- multiplying a row by -1,

- adding an integer multiple of one row to another,

- the analogous column operations, and

- deleting a row and column whose only nonzero entries are a shared 1 on the diagonal (or the reverse of this move).

The canonical form is a diagonal matrix with diagonal entries $d_1, \ldots, d_k$ with $d_i | d_{i+1}$ for $i = 1, \ldots, k-1$ and $d_i \neq 1$ for $i = 1, \ldots, k$. It then follows that

$$G_A \cong \mathbb{Z}_{d_1} \oplus \cdots \oplus \mathbb{Z}_{d_k},$$

where $\mathbb{Z}_0 = \mathbb{Z}$.

These facts are collectively know as the *Fundamental Theorem of Finitely Generated Abelian Groups*. We do not present the formal algorithm for producing the canonical form, but the reader should be able to get the hang of it by working a few examples.

Finally, we note that the order of G_A is given by $|\det A|$ if $\det A \neq 0$ and is infinite if $\det A = 0$. Thus, Theorem 5.1.12 could be restated using the group G_{I-A} and just the sign of $\det(I - A)$.

Exercise 5.1.14 Let $A = \begin{bmatrix} 1 & 2 \\ 2 & 1 \end{bmatrix}$. Show that $G_A \cong \mathbb{Z}_3$.

5.2 Orientation data and stronger invariants

Our strategy for developing more sensitive abstract template invariants is to modify the transition matrix to include orientation information. Given a Markov partition $\{x_1, x_2, \ldots, x_N\}$ of a template we assign an orientation to each partition element. Then the first return map restricted to each partition element is either orientation preserving or orientation reversing.

Definition 5.2.1 A *parity matrix* for a template is constructed from a transition matrix by multiplying a_{ij} by the variable t if the first return map is orientation reversing from the i-th partition element to the j-th partition element.

Example 5.2.2 The matrix $\begin{bmatrix} 1 & 1 \\ 1 & 1 \end{bmatrix}$ is a parity matrix for the Lorenz template, $\mathcal{L}(0,0)$, or, indeed, for any $\mathcal{L}(m,n)$ with m,n even. In contrast, the parity matrix for the horseshoe template $\mathcal{H}$ is $\begin{bmatrix} 1 & 1 \\ t & t \end{bmatrix}$.

In [170] the following theorem is proved:

Theorem 5.2.3 *Let T_1 and T_2 be two abstract templates with parity matrices $A_1(t)$ and $A_2(t)$, respectively. If T_1 and T_2 are related to each other by a finite sequence of template moves then*

$$\det(I - A_1(t)) = \det(I - A_2(t)) \bmod t^2 = 1.$$

Definition 5.2.4 Given a parity matrix $A(t)$, the linear function $\det(I - A(t))$ mod $t^2 = 1$ is the *full Parry-Sullivan invariant*.

The full Parry-Sullivan invariants distinguish the Lorenz template (-1) from the horseshoe template $(-t)$.

The group $G_{I-A(1)}$ is invariant as before, and it is not hard to show that $G_{I-A(-1)}$ is also invariant. It is quite tempting to conjecture that the full Parry-Sullivan invariant, along with these two Abelian groups, would give a complete set of invariants for abstract templates. But the template in Figure 5.3 gives a counterexample. Its full Parry-Sullivan invariant is -1 and both $G_{I-A(1)}$ and $G_{I-A(-1)}$ are trivial, as they are for the Lorenz template. Yet, this template is not orientable and thus clearly inequivalent to the Lorenz template.

Definition 5.2.5 The unit normal bundle of the orbit set of a template is the *ribbon set* of the template. For an embedded template, this set is realized as the bundle of local stable manifolds.

We can reformulate Theorem 5.2.3 in terms of ribbon sets. Let $\mathcal{T}_1$ and $\mathcal{T}_2$ be templates with ribbon sets R_1 and R_2 respectively. Then if there is a homeomorphism between R_1 and R_2 taking ribbons to ribbons (in particular annuli go to annuli, Möbius bands go to to Möbius bands and infinite strips go to infinite strips) and preserving the flow direction, then $\det(I - A_1(t)) = \det(I - A_2(t))$ mod $t^2 = 1$ and $G_{I-A_1(\pm 1)} \cong G_{I-A_2(\pm 1)}$, where $A_1(t)$ and $A_2(t)$ are parity matrices for $\mathcal{T}_1$ and $\mathcal{T}_2$ respectively. Furthermore, the definition of a ribbon set can be extended to basic sets of flows on higher dimensional manifolds and the analogue of these results remain valid.[170] It also follows from [170] that templates with homeomorphic ribbon sets (in the manner just described) can be related, up to embedding, by a finite sequence of template moves.

Definition 5.2.6 Two twist matrices are *flow equivalent* if they are associated with equivalent ribbon sets. The generators of flow equivalence for parity matrices are the analogs of $\overset{s}{\sim}$ and $\overset{e}{\sim}$ for parity matrices, and a new move, the *twist*

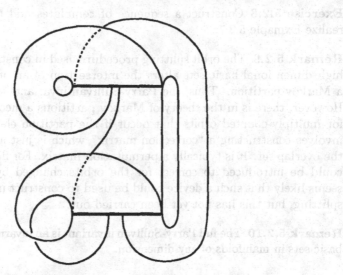

Figure 5.3: A nonorientable template whose full Parry-Sullivan invariant is the same as that of the (orientable) Lorenz template.

move: $A(t) \overset{t}{\sim} B(t)$ if

$$B(t) = \begin{bmatrix} a_{11} & ta_{12} & \cdots & ta_{1n} \\ ta_{21} & a_{22} & \cdots & a_{2n} \\ \vdots & \vdots & & \vdots \\ ta_{n1} & a_{2n} & \cdots & a_{nn} \end{bmatrix},$$

where $A(t) = [a_{ij}]$.

In applying $\overset{t}{\sim}$, we multiply the first row and column of $A(t)$ by t and take $t^2 = 1$. On the level of templates, the twist move corresponds to rotating the bands that pass through the first Markov partition element by a half twist. Thus, among these bands, those which formerly had an odd number of half-twists now have an even number and vice versa. Since this can be realized by isotopy there is no need to define a new corresponding template move.

Example 5.2.7 Let $A(t) = \begin{bmatrix} 0 & 0 & 1 \\ 1 & 1 & 0 \\ t & t & t \end{bmatrix}$, and $B(t) = \begin{bmatrix} 0 & 1 & t \\ 1 & 0 & 0 \\ 0 & 1 & t \end{bmatrix}$. We claim

$A(t)$ and $B(t)$ are flow equivalent. Set $R = \begin{bmatrix} 1 & 1 & 0 \\ 0 & 0 & 1 \end{bmatrix}$, and $S = \begin{bmatrix} 0 & 1 \\ 1 & 0 \\ t & t \end{bmatrix}$.

Now $A(t) = SR$ and $RS = \begin{bmatrix} 1 & 1 \\ t & t \end{bmatrix}$. Applying the twist move followed by an expansion yields $B(t)$.

Exercise 5.2.8 Construct a sequence of templates and template moves that realize Exmaple 5.2.7.

Remark 5.2.9 The orbit splitting procedure used in constructing templates for high dimensional basic sets alters the intersection of an orbit with elements of a Markov partition. Thus, the Parry-Sullivan invariants would suffer changes. However, there is in the theory of Markov partitions a mechanism that corrects for multiply-counted orbits that occur if the partition elements overlap. This involves constructing a "correction matrix" which is just a transition matix for the overlap set. It is typically a permutation matrix. For flows, a similar matrix could be introduced to correct for the orbits changed by orbit splitting. It seems likely that such a device could be used to construct invariants under orbit splitting, but this has not yet been carried out.

Remark 5.2.10 The full Parry-Sullivan invariant is an invariant of one-dimensional basic sets in manifolds of any dimension.

5.2.1 Additional Examples

Example 5.2.11 Figure 5.4 shows two templates each of which has full Parry-Sullivan invariant $-t$. The one on the left has two closed orbits in its boundary while the one on the right has just one such loop; hence, they are distinct. Figure 5.5 shows that the rightmost template is equivalent to the horseshoe template (recall that we are disregarding the embedding).

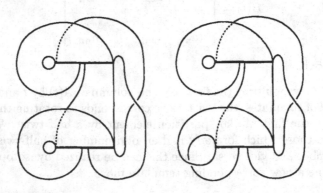

Figure 5.4: Two templates with invariant $-t$.

Example 5.2.12 Consider a template with n strips coming down from a single branch line, each looping back to the branch line and stretching completely across it (while this is not technically a template it is easily turned into one by $n - 2$ small pushes near the branch line: *cf.* the slide move). Suppose that k of the strips are untwisted (orientation preserving) and $l = n - k$ are twisted

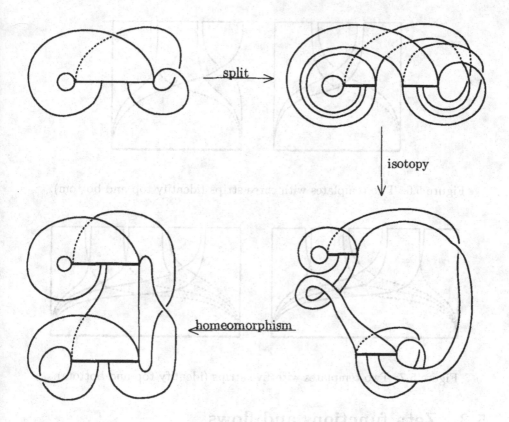

Figure 5.5: A template homeomorphic to the horseshoe template after a split move.

(orientation reversing). Then the full Parry-Sullivan invariant is $1 - k - lt$, and so templates with differing k are distinguished.

Exercise 5.2.13 Show that the Bowen-Franks groups of Theorem 5.1.11 do not further refine the distinctions between the templates in Example 5.2.12

Example 5.2.14 Figure 5.6 shows two templates with three strips, only one of which is twisted in each. They are distinguished by the fact that the number of closed orbits in their respective boundaries differ. In Figure 5.7 we show two templates with five strips, only one of which is twisted in each. A study of the boundary loops, including those with cusps, fails to distinguish them. We conjecture however, that they are distinct and speculate that some type of "non-abelian" invariant is needed to distinguish them.

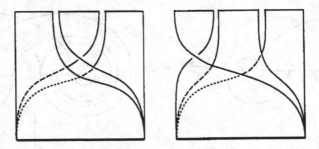

Figure 5.6: Two templates with three strips (identify top and bottom).

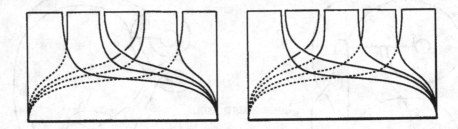

Figure 5.7: Two templates with five strips (identify top and bottom).

5.3   Zeta functions and flows

We now turn to invariants that are sensitive to the embedding of the template. At this stage, knot theory reenters the picture. The idea is again to modify the transition matrix, but this time to produce a *twist matrix*. We shall then use a *zeta function* to count orbits according to the amount of twist in their unit normal bundles. That is, we regard twist as a canonical (though nondynamical) period for a closed orbit in a flow. The weakness of this approach is that invariance holds only over *positive templates*.

5.3.1   Review of Zeta Functions

For general references on zeta functions see [53, Chapter 5] or [162, Chapter 10].

Definition 5.3.1 The *zeta function* of a map $f : M \longrightarrow M$ is the exponential of a formal power series in t,

$$\zeta_f(t) = \exp\left(\sum_{m=1}^{\infty} \frac{1}{m} N_m t^m\right),$$

where N_m is the cardinality of the fixed point set of f^m, the m-th iterate of f.

If f has a hyperbolic chain-recurrent set then the N_m are all finite and $\zeta_f(t)$ is a rational function; hence, a finite set of numbers determine all the N_m. In particular, if O_l denotes the number of periodic orbits of length l then

$$N_m = \sum_{l\mid m} lO_l.$$

We can recover O_l by the Möbius inversion formula [165, page 765]:

$$O_l = \frac{1}{l} \sum_{m\mid l} \mu(m)N_{l/m},$$

where μ is the function defined by

$$\mu(m) = \begin{cases} 1 & \text{if } m = 1, \\ 0 & \text{if } \exists \text{ a prime } p \text{ with } p^2\mid m, \\ (-1)^r & \text{if } m = p_1,\ldots,p_r, \text{ for } r \text{ distinct primes.} \end{cases}$$

When a map f has a zero-dimensional hyperbolic chain-recurrent set, as is the case for subshifts of finite type, then there exists a square matrix A of nonnegative integers such that $N_m = \text{tr}\,(A^m)$. Then $\zeta_f(t) = 1/\det(I - tA)$. The matrix A is of course the transition matrix for a Markov partition.

The difficulty in applying zeta function theory to topological flows is that there is no clear notion of the period of a periodic orbit. Temporal lengths, which are not generally integral, change under reparametrization. On a template, we can use the first return map of a Markov partition to give a (symbolic) period to closed orbits. The zeta function is invariant under the three template moves. However, it is not clear that such an approach would give useful information about the original flow. Instead we use the twist in the local stable manifolds of closed orbits as a canonical period.

Remark 5.3.2 Heuristically, one may view the Parry-Sullivan invariants as the evaluation of a zeta function at ± 1. However, zeta functions typically fail to converge at these values, and the zeta function is not invariant under the expansion move.

5.3.2 Positive Ribbons

A closed ribbon, or ribbon for short, is an embedded annulus or Möbius band in S^3. In this section we define three notions of twist for ribbons. These are, the usual twist τ_u [98, §V], the modified twist τ_m, and the computed twist τ_c.

Like knots and templates, ribbons can be braided. A ribbon which has a braid presentation such that each crossing of one strand over another is positive and each twist in each strand is positive, will be called a positive ribbon. The core and boundary of a positive ribbon are positive braids.

We will use the following notation. If R is a ribbon and $b(R)$ is a braid presentation of R, let c be the sum of the crossing numbers of the core of R, using $+1$ for positive crossings and -1 for negative crossings, as per Figure 1.2. Let t be the sum of the half twists in the strands of $b(R)$ and let n be the number of strands of the core.

Definition 5.3.3 Let $\tau_u = c + t/2$, $\tau_m = n - 1 + t/2$ and $\tau_c = 2n + t$.

Lemma 5.3.4 τ_u *is an isotopy invariant of ribbons over all braid presentations.* τ_m *and* τ_c *are isotopy invariants of positive ribbons over positive braid presentations.*

Proof: For an embedded annulus the linking number of the two boundary components is $c + t/2$. The same formula gives one half the linking number of an embedded Möbius band's boundary with its core. In both cases we find that τ_u is an invariant.

The invariance of τ_m for positive ribbons follows from checking that

$$\tau_m = \tau_u - 2g,$$

where $g = \frac{1}{2}(c - n + 1)$ is the genus of the core of R. Here we have appealed Theorem 1.1.18 for the formula for g. Finally we see that $\tau_c = 2(\tau_m + 1)$. $\square$

For the trefoil orbit in Figure 5.8 the reader can check that $g = 1$ and that its unit normal bundle has $\tau_u = 6$, $\tau_m = 4$ and $\tau_c = 10$.

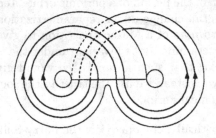

Figure 5.8: Lorenz template with trefoil orbit.

Visually, the conversion of a positive full twist to a loop or writhe decreases t by 2 but creates an extra strand. Since doing this to a negative full twist would increase t by 2 while creating an extra strand, it is easy to show that the invariance of τ_m and τ_c fail for ribbons with mixed crossings. We also note that $\tau_u = \tau_m$ is equivalent to $g = 0$, which in turn is true if and only if the core of the ribbon is unknotted.

Lemma 5.3.5 *For positive templates the number of closed orbits with a given computed twist is finite.*

Proof: Given a positive template we put it into a positive braid form and construct a Markov partition with K partition elements. Given τ_c choose n so that $\tau_c < 2n$. Because the template is braided, a closed orbit that meets any one partition element n times must have wrapped around the braid axis at least n times. Since there are no negative half twists, such an orbit's computed twist is bigger than or equal to $2n$. If $w \geq Kn$, then any closed orbit with symbolic period w must have traveled around the template's braid axis at least n times. Thus, any closed orbit with computed twist τ_c has word length less than Kn. There can only be finitely many such orbits. $\qquad\square$

The computations in the proof of Lemma 5.3.4 show that Lemma 5.3.5 holds for τ_m and τ_u as well as τ_c. This is clear for τ_m. For τ_u, use the fact $g \geq 0$ implies $\tau_u \geq \tau_m$.

5.3.3 Counting Twisted Ribbons

Definition 5.3.6 For a given positive template let $T_{q'}$ be the number of closed orbits with computed twist q'. Let $\mathcal{T}_q = \sum_{q'|q} q' T_{q'}$. Define the *zeta function* of the template to be the exponential of a formal power series:

$$\zeta(t) = \exp\left(\sum_{q=2}^{\infty} \mathcal{T}_q \frac{t^q}{q}\right).$$

Theorem 5.3.7 *The zeta function ζ is an invariant of ambient isotopy of the ribbon set for positive templates. It terms of positive templates ζ is invariant under isotopy and the two templates moves shown in Figure 5.1.*

Proof: This follows directly from Lemma 5.3.4. $\qquad\square$

We now define a *twist matrix*, $A(t)$, whose entries are nonnegative powers of t and 0's, by considering the contribution to τ_c as an orbit goes from one element of a Markov partition to other. Let $A_{ij} = 0$ if there is no branch going from the i-th to the j-th partition element. Let $A_{ij} = t^{q_{ij}}$ if there is such a branch, where q_{ij} is the amount of computed twist an orbit picks up as it travels from the i-th to the j-th partition element. It is easy to see that one can, if necessary, isotope the template so that q_{ij} is always integral. This might be necessary if some of the partition elements lie outside of the branch lines. Also note that one can always choose the partition so that at most one branch goes from the i-th element to the j-th element for each i and j. However, if one wishes to be more general, one can use polynomials in $A(t)$ instead of just powers of t.

For example, the template and partition in Figure 5.9 give

$$A(t) = \begin{bmatrix} 0 & 0 & 0 & t & t \\ 0 & 0 & 0 & 1 & 1 \\ 0 & t^2 & t^2 & 0 & 0 \\ t^2 & t^2 & t^2 & 0 & 0 \\ t^3 & t^3 & t^3 & 0 & 0 \end{bmatrix}.$$

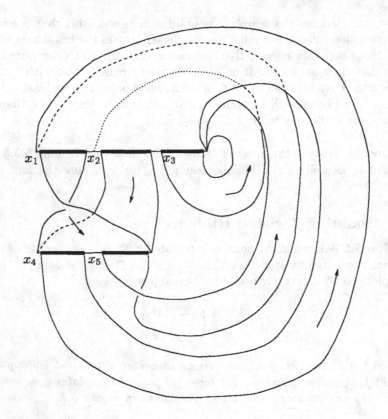

Figure 5.9: A template with a Markov partition indicated by thick lines.

Theorem 5.3.8 *For any template and any allowed choice of $A(t)$ we have* $\zeta(t) = 1/\det(I - A(t))$. *Thus, the zeta function is rational.*

The proof of Theorem 5.3.8 is a standard counting argument and can be found in [171]. We present an example to call attention to the major ideas.

Recall the horseshoe template $\mathcal{H}$ from Figure 2.9. Using the standard two-element Markov partition $\{x_1, x_2\}$, we have

$$A(t) = \begin{bmatrix} t^2 & t^2 \\ t^3 & t^3 \end{bmatrix},$$

and so,

$$1/\det(I - A(t)) = 1/(1 - t^2 - t^3).$$

We apply a standard matrix identity (see Lemma 5.2 of [53] or Proposition 10.7 of [162]) to get

$$\frac{1}{\det(I - A(t))} = \exp\left(\sum_{n=1}^{\infty} \frac{\operatorname{tr} A(t)^n}{n} \right). \tag{5.1}$$

τ_m	0	$\frac{1}{2}$	1	$\frac{3}{2}$	2	$\frac{5}{2}$	3	$\frac{7}{2}$	4	$\frac{9}{2}$	5	$\frac{11}{2}$
$\mathcal{L}$	2	0	1	0	2	0	3	0	6	0	9	0
$\mathcal{H}$	1	1	0	1	0	1	1	1	1	1	1	2
$\mathcal{A}$	3	0	2	0	5	0	10	0	24	0	50	0

Table 5.1: Number of orbits listed by τ_m for different templates.

Let us analyze the first three terms of

$$\sum_{n=1}^{\infty} \frac{\operatorname{tr} A(t)^n}{n} = \frac{t^2 + t^3}{1} + \frac{t^4 + 2t^5 + t^6}{2} + \frac{t^6 + 3t^7 + 3t^8 + t^9}{3} + \cdots$$

There are five closed orbits which pass through the Markov set three or fewer times: x_1, x_2, $x_1 x_2$, $x_1^2 x_2$, and $x_1 x_2^2$. All are unknotted, so $\tau_m = \tau_u$. The t^2 and the t^3 of the first term of the sum correspond to the orbits x_1 and x_2 respectively. In the second term, x_1 and x_2 are counted again, by t^4 and t^6 respectively, since they have been traversed twice. The $2t^5$ corresponds to $x_1 x_2$, where the 2 is the product of number of orbits that pass through the Markov set twice (just 1 in this case) with 2, the number of passes.

The reader should check that $3t^7$ corresponds to $x_1^2 x_2$ and $3t^8$ to $x_1 x_2^2$. The t^6 and the t^9 again count x_1 and x_2 respectively, this time making three trips on each. It is worth noting that $\operatorname{tr}(A(1))^n$ is the number of intersection points of the Markov set with the link of closed orbits which meet the Markov set n' times, where n' divides n.

As a final example, Table 5.1 displays the number of closed orbits having specified (low) amounts of twist for three different positive templates: the Lorenz template, $\mathcal{L}$, the horseshoe template $\mathcal{H}$, and a template denoted $\mathcal{A}$, shown in Figure 5.10. The template $\mathcal{A}$ was first studied in [169], where it was shown to contain only prime knots.

Exercise 5.3.9 Write a computer program to generate table entries similar to Table 5.1 where the user enters the twist matrix.

Remark 5.3.10 Using zeta functions to count twists is a strategy which cannot be adapted to all templates. Recall the templates $\mathcal{U}$ and $\mathcal{V}$ from Chapter 3; since there exist isotopic template renormalizations on these templates, each contains infinitely many distinct copies of a knot with a given twist.

5.4 A zeta function for Lorenz attractors

Branched 2-manifolds with semiflows were first introduced to study the strange attractors believed to be associated with the Lorenz equation (Equation (2.1)) [193], [194]. Since the hyperbolicity of the Lorenz equations in the parameter

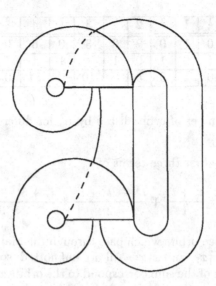

Figure 5.10: The template $\mathcal{A}$.

range of interest was and still is unknown, geometrically defined flows were used as a model. The attractors of the model flows could then be studied rigorously via templates. See [166, Appendix G] for a nice overview.

However, these "early" templates differ in two respects from the Lorenz template $\mathcal{L}(0,0)$ defined in Chapter 2, and indeed, from all of the templates discussed so far. First, orbits in the boundary can enter the interior of the template — that is, the boundary flow is not invariant. In particular, the closed orbits x_1^∞ and x_2^∞ are not realized. Secondly, the template includes a saddle point, O. This causes the invariant set of the template to be two dimensional. Figure 5.11 shows this object, which we shall call a *sublorenz template* can be used to model a geometric Lorenz attractor. Although this is not a subtemplate of the Lorenz template $\mathcal{L}(0,0)$, all of the closed orbits on it are ambient isotopic to knots in the Lorenz template. As before, we may use words in x_1 and x_2 to describe orbits; however, since we will work only with templates having two elements in the Markov partition, *we will relabel x_1, x_2 as x and y respectively for the remainder of this section.* Note in addition that the line we use for a cross section of the semiflow extends beyond the branch set. We shall call it the *extended branch line*.

Consider the saddle point within the sublorenz template. On this template (and in the full three-dimensional flow which generated it), the saddle point and the attractor are inseparable but distinct invariant sets. Thus, the Lorenz attractor is not closed: *cf.* Theorem 1.2.13. Of special interest are the trajectories of the left and right branches of $W^u(0)$. Denote these l and r respectively. If they each return to 0, thus forming a double saddle connection, we can define a

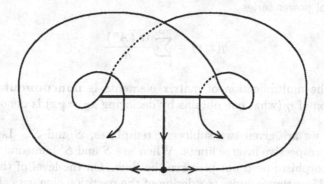

Figure 5.11: A sublorenz template.

finite Markov partition for the semiflow: see Figure 5.12. This naturally leads to a corresponding transition matrix $A(x, y)$ which measures not only which partition element sequences are admissible, but also along which strip (x denoting left and y denoting right) the transitions occur (see Example 5.4.4 below). Although the double saddle connection case is not a generic case, it is the situation we consider.

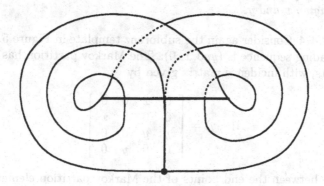

Figure 5.12: A double saddle connection.

Two tools allow us to compactly encode information on the transitions in a sublorenz template.

Definition 5.4.1 The *kneading sequence* **k** of a sublorenz template is a pair of sequences $(\mathbf{k}_l, \mathbf{k}_r)$ defined as follows: $\mathbf{k}_l$ is a sequence of x's and y's determined by the order in which l meets the extended branch line. If l returns to the saddle point then a terminal 0 is appended to $\mathbf{k}_l$. The sequence $\mathbf{k}_r$ is defined similarly.

Definition 5.4.2 Let S denote a sublorenz template with finite kneading sequence and transition matirx $A(x, y)$. Then the *pre-zeta function* of S is defined by the formal power series

$$\eta(x,y) = \sum_{i=n}^{\infty} \frac{\text{tr}\,(A^n)}{n}. \tag{5.2}$$

Note that the multiplication of matrix elements is **noncommutative**. The abelianization of η (what one obtains by declaring $xy = yx$) is denoted η^a.

Suppose we are given two sublorenz templates, S and S'. Let $\hat{S}$ and $\hat{S}'$ denote their respective inverse limits. When are $\hat{S}$ and $\hat{S}'$ homeomorphic? Here the homeomorphism need not preserve the flow. On the level of the templates we only need invariance under reordering of the partition elements. In [194], two answers are given via the previous two definitions.

Theorem 5.4.3 (Williams [194]) *Let $\mathcal{L}$ and $\mathcal{L}'$ denote sublorenz templates with finite kneading sequences. Then the following statements are equivalent:*

(a) *$\mathcal{L}$ and $\mathcal{L}'$ have homeomorphic inverse limits;*

(b) *The corresponding kneading sequences are equal, $\mathbf{k} = \mathbf{k}'$; i.e., $\mathbf{k}_r = \mathbf{k}'_l$ and $\mathbf{k}_l = \mathbf{k}'_r$; and*

(c) *The corresponding pre-zeta functions are equal, $\eta(x,y) = \eta'(x,y)$, up to exchanging x and y.*

Example 5.4.4 Consider again the sublorenz template in Figure 5.12, denoted S. The kneading sequence is $(yy0, xx0)$. The Markov partition has the obvious four elements, with incidence matrix given by

$$A(x,y) = \begin{bmatrix} 0 & x & 0 & 0 \\ 0 & 0 & x & x \\ y & y & 0 & 0 \\ 0 & 0 & y & 0 \end{bmatrix}.$$

The overlap between the end points of the Markov partition elements does not cause any over counting problems since the end points all flow towards the saddle point 0 and so are not periodic. The abelianized pre-zeta function is then determined by

$$\exp(-\eta^a(x,y)) = \det(I - A) = 1 - xy - xy^2 - x^2y - x^2y^2.$$

That is, after abelianization the usual tools of zeta function theory can be applied. But it is not clear how to define a non-abelian zeta function using a matrix formula. One apparently has to grind out the trace of each power of the matrix directly. For the matrix $A(x,y)$ the first three terms of η are

$$\frac{0}{1} + \frac{xy + yx}{2} + \frac{x^2y + xyx + xy^2 + yx^2 + yxy + y^2x}{3}.$$

As orbits xy and yx are the same, abelianization would not cause any loss of invariant information in the second term. Likewise the elements of the third term correctly capture the two period three orbits. This is because abelianization and cyclic permutation are the same for these two terms. But, by the fifth term this is no longer the case. The reader can check that there are no orbits with the word x^3y^2 on S, but the word x^2yxy is realized by a trefoil orbit. This distinction is lost in η^a but not by η.

In [196] Williams developed a new type of determinant that allows one to write a matrix equation analogous to Equation (5.1). We give a heuristic outline and an example.

Given a Markov partition with n elements consider the set of closed orbits which do not visit any partition element more than once. These orbits all have (symbolic) period less than or equal to n. For the template S they are xxy, $xxyy$, xy, xyy. Each orbit corresponds to a cyclic permutation class in the free group on two symbols. Following [196] we call these classes *free knot symbols*. For S the free knot symbols are just (xxy), $(xxyy)$, (xy), and (xyy), where the parentheses denote the cyclic permutation class. We allow, for algebraic reasons, the empty symbol (). Next, we define a *free link symbol* as a formal product of free knot symbols whose corresponding knots have no partition elements in common, where the empty symbol () is taken to be the unit. We will consider the ring of free link symbols given by allowing formal addition of symbols with integer coefficients. For the template S, each free link symbol is the product of just one free knot symbol.

Given any square matrix A of x's, y's and 0's one can write down all the free link symbols. To do this we first define an *index cycle*. An index cycle is a finite sequence, $(i_1, \ldots, i_k)$ of k distinct integers, $0 \le k \le n$ such that the product of matrix elements

$$A_{i_1,i_2} A_{i_2,i_3} \cdots A_{i_k,i_1} \ne 0.$$

Then

$$(A_{i_1,i_2}, A_{i_2,i_3}, \ldots, A_{i_k,i_1})$$

is a free knot symbol for the incidence matrix. The empty symbol is corresponds to an empty index cycle: this is the multiplicative identity in the ring. We may then concatenate free knot symbols so long as their corresponding index cycles have no common elements. This yields the collection of free link symbols for A, denoted $fls(A)$.

We make the following observations. The free knot symbols (xy) and (yx) are the same by cyclic permutation. But $(xxyyx)$ is different from $(xyxyx)$. This is as it should be to model knots on a template. However, the ring product is commutative. Again this makes sense, since there is no preferred order on the link of periodic orbits. Thus in the definition below $(w)(v)$ and $(v)(w)$ represent the same element of the ring. Ring addition is also (of course) commutative. The addition operation should thought of as "purely algebraic", in that unlike the ring product it does not correspond to a geometric operation on knots.

Definition 5.4.5 The *link-determinant* is defined by

$$\text{link-det } (I - A) = \sum_{fls(A)} \sigma_i \mathbf{w}_i, \qquad (5.3)$$

where $\mathbf{w}_i = (w_1) \cdots (w_l) \in fls(A)$ and $\sigma_i = (-1)^l$. For the template $\mathcal{L}$ we get $1 - (xxy) - (xxyy) - (xy) - (xyy)$ as the link-det of the incidence matrix. The $(I - A)$ in the above definition may look a bit odd at first. It can be regarded as a notational formality for consistency with the usual zeta function. However, allowing 1's in the matrix can be used to give a definition of free link symbols so as to have them all be of length n by "filling" in with 1's. See [196] and [103].

Exercise 5.4.6 Let

$$A = \begin{bmatrix} 0 & x & 0 \\ 0 & 0 & x \\ y & y & 0 \end{bmatrix}.$$

Show that link-det $(I - A) = 1 - (xxy) - (xy)$.

Theorem 5.4.7 (Williams [196]) $\exp(-\eta(x, y)) = \text{link-det } (I - A)$.

The intuitive idea is that *most* of the non-abelian "badness" is "hidden" inside the free knot symbols and so one can use standard matrix theory machinery, suitably modified. In particular an analogue of the Cayley-Hamilton theorem holds [103]. To see why we say most and not all of the non-abelian badness is hidden, see Example D of [196]. We name $\zeta_W(x, y) = \exp(-\eta(x, y))$, the *Williams zeta function*.

Theorem 5.4.7 can be interpreted to mean that a small set of words, corresponding to links "fitted" to a Markov partition, determine all the other possible periodic words of the given Lorenz attractor. Since the order of the words has not been washed out by abelianization, we can reconstruct the knots. This is not too surprising since the kneading sequence can be viewed as two special knots that determine all the others. In fact, the words corresponding to the two knots $l \cup O$ and $r \cup O$, do appear in the link-det.

Finally, we note that under abelianization link-det $(I-A)$ becomes det $(I-A)$ and that if P is a permutation matrix link-det $(I - A) = \text{link-det } (I - PAP^{-1})$. These facts are both have easy proofs and are done in [196].

Example 5.4.8 Figures 5.13 and 5.14 show two sublorenz templates, $\mathcal{A}$ and $\mathcal{B}$. It is not hard to set up the corresponding matrices $A(x, y)$ and $B(x, y)$ and compute that

$$\det(I - A) = x^9 y^6 + x^8 y^5 + x^7 y^5 - x^6 y^4 - x^3 y^2 - x^2 y + 1 = \det(I - B).$$

However, $\mathcal{A}$ and $\mathcal{B}$ are not equivalent as can be seen by checking their kneading sequences. We leave it as an exercise to compute their Williams zeta functions.

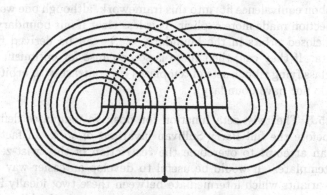

Figure 5.13: The sublorenz template $\mathcal{A}$.

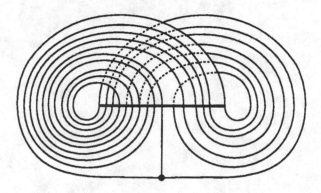

Figure 5.14: The sublorenz templates $\mathcal{B}$.

5.5 Remarks on other invariants and open problems

Remark 5.5.1 A new class of template invariants has recently been announced [100]. They are derived from em quantum groups, a class of objects which appears to be of fundamental importance in the study of knot and link invariants [158]. These results are beyond the scope of the present text, but it is worth noting that both the original Parry-Sullivan invariant and the full Parry-Sullivan invariant have been realized as quantum invariants. However, the computations involved in developing more sensitive invariants with regard to embeddings appear to be quite hard and still remain to be done.

Remark 5.5.2 There has been a great deal of work in symbolic dynamics of subshifts of finite type under various restrictions (e. g. irreducibilty) and in generalized contexts (e. g. finite identifications). See [32], for example. Our notion of ribbon equivalence fits into this framework, although one would hope to see the connection made more explicit. It is less clear if our boundary invariants (the link of closed orbits in the boundary, etc.) can be derived from purely symbolic data. If they can, then there is some cause for optimism toward the problem of classifying ribbon sets of templates (in contrast with arbitrary ribbon sets, which do not have boundary).

Remark 5.5.3 The twist-zeta function for positive templates defined in §5.3 was found before the full Parry-Sullivan invariant of §5.2. In fact, the latter arose from an attempt to overcome the restriction of the twist-zeta function to positive templates. It would be useful to develop an easier way to compute template invariants which intermediate between these two; ideally it should be well-defined for *all* templates but should contain more embedding information than does the full Parry-Sullivan invariant.

Chapter 6: Concluding Remarks

In this monograph we have described tools, developed largely in the past fifteen years, which permit the explicit construction and description of those knot and link types realised as periodic orbits in certain classes of three-dimensional flows. The principal tool is the template, which allows the reduction of a three-dimensional flow having a hyperbolic invariant set to a semiflow on a branched two-manifold. We also develop a "template calculus:" a symbolic language for the characterization and manipulation of templates. These techniques are described in Chapter 2. They build on "classical" ideas from knot theory and dynamical systems theory, which we review in Chapter 1.

In Chapter 3 we have used these tools to derive general results on template knots, and to prove the existence of a universal template which contains (infinitely many) representatives of all tame knots and links. Here the tone is that of inclusion. Chapter 4 takes a more exclusive viewpoint; we focus on restricted classes of templates, especially that corresponding to the "simplest" suspension of Smale's horseshoe map. We show that in such cases only limited classes of knots can occur, and that uniqueness results may be used to distinguish branches of periodic orbits in bifurcation studies. The chapter ends with a return to inclusiveness, as we show that the universal template of Chapter 3 occurs within the flows of an open set of ODEs near a double Silnikov type homoclinic bifurcation point.

Chapter 5 takes a different direction in that we turn to the characterization of templates *per se* instead of the knots and links they support. Template invariant theory is less well-developed than the corresponding theory for knots, and this chapter is necessarily more tentative in nature and limited in scope than the rest of the book.

In the course of the text we have noted or hinted at a number of open questions. In the hope that they may stimulate future work, we collect and expand on them here. We also give references to some relevant (and mostly recent) literature of which we learned shortly before the book went to press.

Problems in template theory and applications

Problem 6.0.1 The best sort of result one could hope for in template theory would be an easily-computed, discriminating template invariant. This appears to be a very difficult undertaking, as mentioned in Chapter 5. However, as the number of new knot-and-link invariants seems to be growing daily, there is hope that some of these recent invariants can be exported to template theory: *e.g.*, the quantum template invariants mentioned in §5.5.

Problem 6.0.2 As an alternate approach to the previous problem, it would be very useful (and indeed, it seems quite feasible) to develop a rough classification theorem for templates. The crudest such result would provide necessary and sufficient conditions for determining when a template is universal. Natural refinements of this classification would include a compact way to describe how a template fails to be universal (*e.g.*, the template is positive). Since we have shown that every template is universal up to embedding, this would entail some sort of description of how the strips are embedded (*e.g.*, they are all linked in too-complicated a manner, or perhaps each strip is knotted and forces satellite knots, etc.). We recall Conjecture 3.2.24, which states that a template is universal if it has a sufficiently large unlink within it — failure to be universal may be encoded in the size of the largest unlink. A related problem is to determine whether or not a universal template (one which contains all knots) must be *very universal* in the sense that it contains $\mathcal{V}$ as a subtemplate (and hence, all links, infinitely many copies of all links, etc.). However, this appears to be a rather messy problem.

Problem 6.0.3 There are several lesser problems concerning universal templates. For example, how are the knot types distributed in the space of periodic orbits? Are the unknots dense in this space? Answers to such questions would give an idea of the probability of finding a particular type of knot within the periodic orbit set.

Problem 6.0.4 In applying template theory to studying fibred knots (recall §2.3.4) it is unclear how much information is encoded in the template associated to the fibration. In all the examples computed here (related to the figure-eight knot and the Whitehead link), the derived templates are universal. It is reasonable to guess that every fibred link with pseudo-Anosov monodromy which is not a positive braid has a universal template associated to its fibration. However, if this is not true, then the templates would serve as a tool for distinguishing certain fibred links. Or, perhaps, finer information than the planetary link as a whole could be derived from the template.

Problem 6.0.5 In applying template theory to templates derived from flows, we have restricted ourselves to uniformly hyperbolic dynamical systems, for which the Template Theorem applies. It would be of great interest to adapt the proof to non-uniformly hyperbolic cases (covered by Pesin theory), which are known to be crucial for describing the full dynamics of smooth maps of Hénon type and their attractors [131, 140].

Problem 6.0.6 In a related vein, the material of Section 5.4 also suggests a new direction. Indeed, while the study of templates for hyperbolic sets has matured over the past fifteen years, there have been few application of templates to *attractors per se*. This is perhaps mainly because it is very difficult to prove that non-trivial, indecomposable attractors exist for flows defined by specific ODEs, while hyperbolic (sub-) sets are relatively easy to find. We note that Kennedy in his Ph. D. dissertation [102] shows that the Lorenz-like templates (Section 2.3.1) are realized as models for attractors in certain geometrically defined flows, and there has been some interesting work showing that certain classes of ODEs contain geometric Lorenz attractors: see [44], [156], and [152]. However, no other type of template has been rigorously associated with the full attractor of an ODE. The examples given in Section 2.3.3, and the proof of Section 4.4.2 that a universal template lies in the flow near a double Silnikov homoclinic connection, all involve hyperbolic sets which may *belong to* an attracting set, but which certainly do not comprise the whole attracting set.

A further complicating factor, mentioned briefly in Section 5.4, is the issue of invariant sets or attractors with infinite (countable) Markov partitions, which may require kneading theory for a full description, as does the (geometrical) Lorenz attractor. Williams [194] gives a method for the construction of infinite Markov partitions for the sub-Lorenz templates of Chapter 5. J. Wagoner [185], [186], has also studied infinite Markov partitions, but not in the context of templates. This area is also open.

Problem 6.0.7 The largely non-rigorous ideas of Section 2.3.5, in which templates are derived from embedded (experimental) time series, continue to attract interest. Papers following up on [128] include [126, 121] and [108, 159, 109, 111, 110, 113, 112]. The reference [126] is notable in that it shows explicitly how different embeddings can give rise to templates carrying topologically distint links of periodic orbits (although this is not surprising, in view of the fact that all templates are universal, up to embedding (Theorem 3.3.5).) It would thus seem important to derive embedding-invariant descriptions of templates, *cf.* the Parry-Sullivan invariants of Chapter 5.

Problem 6.0.8 Perhaps the greatest shortcoming of the techniques detailed in this book (except for portions of Chapter 5) is their inherent three-dimensionality. Knotting and linking of periodic orbits is simply impossible in higher dimensions. In terms of trying to derive topological information from time series data, [136] and [127] are good first steps in deriving higher dimensional topological structures from time series.

Other avenues are also open. There is a well-defined notion of higher-dimensional knot theory in which k-spheres are knotted and linked within $(k+2)$-spheres. Several authors have suggested applying such perspectives to dynamical problems [128, 130]; however, there is a glaring lack of dynamically relevant spheres except for 1-spheres (periodic orbits). What *can* (and should) be explored is the presence of knotted k-tori in $(k + 2)$-dimensional flows. Such tori may be nontrivially knotted, thought not in the way that one might expect, given one's intuition in $\mathbb{R}^3$. Here is an example: consider a nontrivial knot $K \subset \mathbb{R}^3$. Then $K \times S^1 \subset \mathbb{R}^3 \times S^1$ is a nontrivially knotted torus in a 4-manifold. It is clear to see how such knotted tori would arise naturally in several contexts, including periodically excitation of three dimensional ODEs possessing hyperbolic periodic orbits.

In this context it remains to develop a good knot theory for embedded tori (almost all of the work in higher-dimensional knot theory has been done with spheres), and then to find key examples in which embedding information can be easily derived. It appears unlikely that a higher-dimensional template theory is possible; however, considering the embedding data in Hamiltonian systems might be a good place to start.

Appendix A: Morse-Smale / Smale Flows

A.1   Morse-Smale flows

In Morse-Smale flows the basic sets are simply closed orbits and fixed points: there is no "chaos" and hence little need for templates. Nevertheless, such flows form an interesting and important class. Here we review basic facts about Morse-Smale flows, culminating in the result of M. Wada [184] that characterizes which links can be realized as the periodic orbit link of a nonsingular Morse-Smale (NMS) flow on the 3-sphere. (Recall that a nonsingular flow is a flow without fixed points.) Surprisingly, a subclass of these links is precisely the set of realizable links in a special class of Hamiltonian systems [35] (see Remark A.1.14).

We recall the definition of Morse-Smale flows from Chapter 1:

Definition A.1.1 A flow ϕ_t on a manifold M is *Morse-Smale* if,

- The chain recurrent set is hyperbolic,

- The stable and unstable manifolds of basic sets meet transversely.

- Each basic set consists of a single closed orbit or fixed point.

For M a compact manifold, it follows that there are a finite number of periodic orbits and fixed points.

Among structurally stable flows, Morse-Smale flows have attracted special interest. Morse-Smale flows are dense in the C^1 topology of C^1 flows on compact 2-manifolds (this follows from Pugh's closing lemma [147]). In the C^∞ case the density result is known only for orientable compact 2-manifolds [142] and for the projective plane, the Klein bottle or the torus with a cross cap [78]. For other nonorientable 2-manifolds the question remains open. On any manifold, Morse-Smale flows form a dense subset among the gradient flows, regardless of the smoothness class. An excellent account of these results can be found in [139] and the references there.

Example A.1.2 We give a construction for a NMS flow on S^3 with two closed orbits: one attractor and one repellor, which form a Hopf link as illustrated in Figure 1.9(c). Consider the solid torus $V_1 = D^2 \times S^1$ as the subset of $\mathbb{R}^2$ (in polar coordinates) crossed with S^1 given by

$$V_1 = \{(r, \theta, \phi); 0 \leq r \leq 1, \theta, \phi \in S^1\}.$$

Place a flow on V_1 given by the vector field

$$X = (\dot{r}, \dot{\theta}, \dot{\phi}) = (-r, 0, f(r)),$$

where $f(r)$ is a smooth nonnegative bump function with support in a small neighborhood of $r = 0$. Let V_2 denote a second copy of V_1 outfitted with the "backwards" vector field $-X$. As such, we may match the vector fields on the boundaries of V_1 and V_2 and glue these solid tori together via $\Phi : \partial V_1 \to \partial V_2$ given by $(\theta, \phi) \mapsto (\phi, \theta)$.

There are several ways to show that gluing V_1 and V_2 together in this manner yields S^3, concluding the existence of the desired NMS flow: we review one such procedure. Observe that gluing two disks together along their boundary in the obvious way produces a 2-sphere. Likewise gluing two 3-balls together yields a 3-sphere. If we cut out a small neighborhood of a diameter in one of the 3-balls, the remaining portion of that 3-ball is a topological solid torus. However, the union of this neighborhood and the other 3-ball is also a solid torus. Thus, we have realized S^3 as a union of two solid tori (in this case, V_1 and V_2) glued together along their boundaries in a manner which exchanges the meridian and longitude as per Φ. The resulting NMS flow is pictured in Figure A.1.

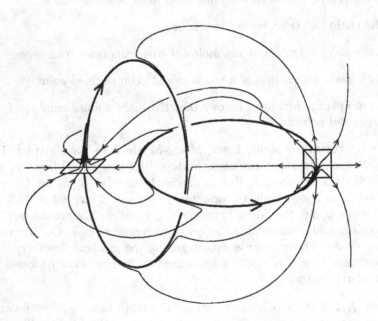

Figure A.1: A NMS flow on S^3 which has one attractor and one repellor arranged in a Hopf link.

Not every manifold supports a nonsingular Morse-Smale flow, or even a nonsingular flow for that matter. A simple Euler characteristic criterion determines if a manifold supports a nonsingular flow, Morse-Smale or otherwise. This criterion is a mild extension of classical results due to H. Hopf and Poincaré [124]:

Lemma A.1.3 *Let M be a compact manifold whose boundary, possibly empty, has been partitioned into two collections of connected components, $\partial_- M$ and $\partial_+ M$:*

$$\partial M = \partial_- M \cup \partial_+ M,$$

$$\emptyset = \partial_- M \cap \partial_+ M.$$

Then there exists a nonsingular vector field on M, pointing inward on $\partial_- M$ and outward on $\partial_+ M$, if and only if $\chi(\partial_- M) = \chi(M)$. [1]

Asimov [12] has shown that every manifold of dimension $n \neq 3$ which satisfies the Euler criterion above supports a nonsingular Morse-Smale flow. This is false for 3-manifolds, but Morgan [132] has characterized which 3-manifolds support nonsingular Morse-Smale flows. Morgan's criteria are rather technical and we will not go into them here. See [132] or [35]. The basic idea behind these results is that a manifold supports a nonsingular Morse-Smale flow if and only if it admits a *round handle decomposition*. We give details only for the case of 3-manifolds.

A.1.1 Round handles

In dimension three, a round handle (RH) is a solid torus $D^2 \times S^1$ together with a specified subset of its boundary called its *attaching zone*. We imagine that each round handle comes with a NMS flow having the core $\{0\} \times S^1$ as the sole closed orbit, as in Example A.1.2. The exit set of the flow will be the attaching zone for the round handle (possibly empty, in the case of attracting orbits). We will use round handles to build NMS flows by gluing them together so that the attaching zones are joined to the in-flowing regions of other round handles.

- **0-RH:** The attaching zone is the empty set and the core is an attracting orbit. We start building a NMS flow by laying down some 0-RHs.

- **1-RH (untwisted):** The attaching zone consists of two disjoint annuli, each going longitudinally around the torus once, and the core orbit is a saddle orbit whose local stable and unstable manifolds are annuli (perhaps twisted with a nonzero but even number of half twists).

- **1-RH (twisted):** The attaching zone is an annulus that wraps twice longitudinally about the torus, and the core orbit is a saddle whose local stable and unstable manifolds are Möbius bands.

- **2-RH:** The attaching zone is the entire boundary, and the core orbit is a repellor.

---

[1] Recall $\chi(\emptyset) = 0$. For review of the Euler characteristic, see [117].

Remark A.1.4 This definition can easily be extended to define round handles in higher dimensions: see [12].

Definition A.1.5 A RH decomposition of S^3 is a sequence of manifolds:

$$\emptyset = M_0 \subset M_1 \subset M_2 \subset \cdots \subset M_k = S^3$$

such that each M_j is obtained by attaching a RH to M_{j-1} along its attaching zone.

Lemma A.1.6 (Asimov [12] and Morgan [132]) *For every RH decomposition of S^3 there is a NMS flow on S^3 such that (1) the closed orbits of the flow are equivalent to the cores of the round handles, together with their indices and twistedness; and (2) the flow is inwardly transverse to ∂M_j for each j.*

Conversely, for every NMS flow on S^3 there is a RH decomposition such that (1) and (2) above hold.

Sketch of Proof: It is clear from the remarks above that if we can find a round handle decomposition, then we can build a corresponding NMS. One does have to check that the stable and unstable manifolds intersect transversely, but this can always be achieved by a small perturbation.

The other direction is harder and will require the use of the *no-cycle property* of Morse-Smale flows. Since in a NMS flow, all the closed orbits are attractors, repellors, or saddles, their tubular neighborhoods are round handles. We want to use the action of the flow itself to do the attaching. But we need to order the orbits sequentially to get a decomposition. In our case, we would like to enumerate all the attracting orbits in arbitrary order, then the saddles, and finally the repellors, again in any order; however, the saddles cannot be attached in arbitrary fashion. Clearly, if the unstable manifold of one orbit flows into the stable manifold of another, this latter orbit should appear first in the decomposition. But should the unstable manifold of this orbit flow back into the stable manifold of the former, a decomposition would not exist. It is the *no-cycle property* which circumvents this problem.

Let $c_1, \ldots, c_n$ be the closed orbits of a NMS flow. Define $c_i \leq c_j$ if the unstable manifold of c_j meets the stable manifold of c_i. The No-Cycle Theorem [165] states that $\leq$ is a partial ordering on the closed orbits. By choosing any total ordering compatible with $\leq$, we may use the action of the flow to attach tubular neighborhoods of the closed orbits and obtain a decomposition.

Suppose we have built up M_{i-1}, and want to attach the next round handle. (M_0 is easy as it is just a 0-RH.) Let N_i denote the neighborhood of c_i and let E_i denote the exit set of the flow. The forward image of E_i under the flow intersects ∂M_{i-1}. We form a bigger round handle by joining N_i with $\bigcup_{t>0} \phi_t(E_i)$ and deleting any intersection with M_{i-1}. Taking the closure of this yields a RH for c_i attached to M_{i-1}. A small adjustment must be applied to the boundary of M_i, which is tangent to the flow along the "edges" of $\bigcup_{t \geq 0} \phi_t(E)$. In addition, one must also adjust slightly to make sure things are smooth.   □

A.1.2 The 3-sphere

In this book, we have considered the knotting and linking properties of closed orbits for flows on the 3-sphere. In [184], M. Wada characterized the class of links that could be realized as the set of periodic orbits of a nonsingular Morse-Smale flow on S^3. Actually he does a little more — each component of a link of closed orbits may be labeled with the index of the orbit: 0 (for attractors), 1 (saddles) or 2 (repellors). Wada characterizes which *indexed links* can be realized.

The interested reader may find Wada's paper tersely written. In particular, there are no illustrations, although the proof requires nontrivial visualization.[2] A more recent paper [35] (see Remark A.1.14) is easier to follow, but leaves out some details, referring to Wada's paper. Thus, the diligent reader might want to have both papers on hand to understand the proof. Here we present only a statement of the result and a brief outline of the proof. Before stating Wada's theorem, we construct two further examples of NMS flows on S^3. Each example shows how to build a new flow from one or more existing flows.

Example A.1.7 Consider an attractor A of a NMS flow on S^3. We may remove a tubular neighborhood N of A and replace it with a solid torus supporting an NMS flow which is inwardly transverse to the boudary, but which contains more than a single closed orbit. Consider the return map on a meridional cross-section of N: this will appear as a disc with a sink at the center of the disc, the remainder of which is foliated by invariant radial lines along which orbits tend towards the sink.

In Figure A.2, we give three different examples of new flows that can be glued in to $S^3 \setminus N$, illustrated by means of the cross-sectional return maps. Note that each has three closed orbits (or fixed points in the map), and that one is a saddle (as should be via simple index theory). Upon suspension of these maps, the two "side" orbits may cable about the core orbit an arbitrary number of times. Finally, we may generate all sorts of variations on this example by performing an n-fold branched covering of the disc, branched over the center point, as illustrated in Figure A.3 — hence, more general cablings of orbits can be produced. Of course, one may reverse the flow direction and create NMS flows on solid tori with the attractors and repellors exchanged and the flow outward on the boundary.

We now possess several tools and components for building new NMS flows on S^3 from old ones. We next construct a NMS flow on S^3 with basic sets consisting of a single saddle orbit and two Hopf links, each a repellor-attractor pair, put together via a "split sum."

Definition A.1.8 (Split sum) Let L_1 and L_2 be links in two three-spheres S_1^3 and S_2^3 respectively. Delete a small open 3-ball from each of the link complements, $S_i^3 - L_i$, $i = 1, 2$, and form the union of $S_1^3 - B_1$ and $S_2^3 - B_2$ by gluing them along their boundaries. We obtain a new 3-sphere (to see this take one of

---

[2] A preprint of Wada's paper did include many helpful illustrations which did not survive in the published version.

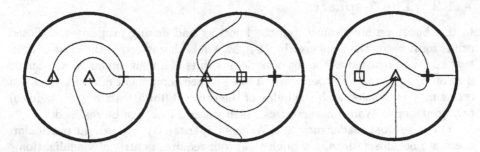

Figure A.2: Return maps on a cross section of an attracting orbit. Triangles refer to sinks, squares to sources, and crosses to saddles.

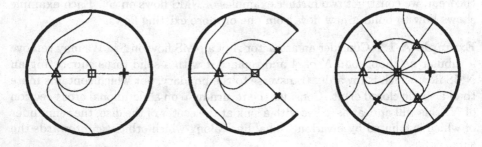

Figure A.3: Cablings more general than $(2, n)$ may be created by modifying one of the above examples via a branched covering.

the balls to be a neighborhood of "∞") with a new link denoted $L_1 \circ L_2$ and called the *split sum* of L_1 and L_2.

Taking the split sum of two links results in a separable link.

Example A.1.9 We will build up our flow in pieces and then glue the pieces together to obtain a flow on S^3. Let C denote a cylinder $I \times S^1$. We can put a NMS flow on the thick cylinder $C \times I$ having a single closed orbit of index one, *i.e.*, a saddle: see Figure A.4. The exit set is $\partial C \times \text{int} (I)$. The flow enters from $\text{int}(C) \times \partial I$ and is transverse along the exit and entrance sets. The saddle orbit is the center circle of C cross the midpoint of I.

Definition A.1.10 A simple closed curve embedded in a surface is *inessential* if it bounds a disk in the surface. Otherwise, the curve is said to be *essential*.

Now we continue with Example A.1.9. Let V_i, $i = 1, 2$ be two 0-round handles. Attach one component of $\partial C \times I$ to an inessential annulus on ∂V_1, so that the annulus' core bounds a disk in ∂V_1, and attach the other component to an inessential annulus on ∂V_2. We can "round off the corners" of this attaching

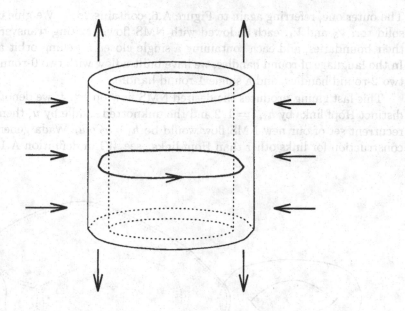

Figure A.4: Thickened cylinder with a saddle orbit.

so as to obtain a smooth flow on the union with the flow entering transversely along the entire boundary of the resulting manifold.

However, if we attach $V_1, C \times I$, and V_2 *naively* as in Figure A.5 there would be a 2-sphere transverse to the flow in the boundary. Any attempt to use this to build a flow on S^3 would force a singularity. Thus the attachment to V_2 must be done in a different way. In Figure A.6, $C \times I$ "swallows" V_2, and then turns in to attach to it. Note that $\partial C \times \{0\}$ bounds a disk in V_1 minus the attaching annulus but not on V_2 minus the attaching annulus.

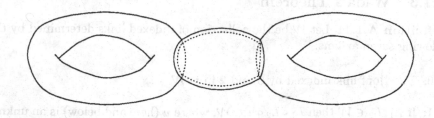

Figure A.5: The union of two solid tori and a thick cylinder may have a sphere and a double torus as boundary.

To recap so far, the manifold $V_1 \cup (C \times I) \cup V_2$ has a NMS flow with three closed orbits: two attractors and a saddle. The flow is transverse inward along the entire boundary. What is that boundary? It is the disjoint union of two tori.

The outer one, referring again to Figure A.6, contains "∞". We glue in two new solid tori V_3 and V_4, each endowed with NMS flows, exiting transversely along their boundaries, and each containing a single closed repelling orbit at its core. In the language of round handles, we have built a flow with two 0-round handles, two 2-round handles, and a single 1-round handle.

This last gluing produces the desired NMS flow on S^3. If we denote a pair of distinct Hopf links by h_i, $i = 1, 2$ and the unknotted saddle by u, then the chain recurrent set of our new NMS flow would be $h_1 \circ h_2 \circ u$. Wada generalizes this construction for links other than Hopf links: see *W1* in definition A.1.11 below.

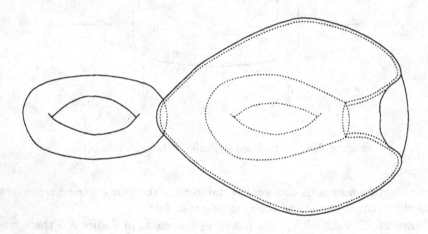

Figure A.6: The same handles attached differently contain only tori as boundary components.

A.1.3   Wada's Theorem

Definition A.1.11 Let $\mathcal{W}$ be the collection of indexed links determined by the following seven axioms:

W0: The Hopf link indexed by 0 and 2 in is $\mathcal{W}$.

W1: If $L_1, L_2 \in \mathcal{W}$ then $L_1 \circ L_2 \circ u \in \mathcal{W}$, where u (here and below) is an unknot in S^3 indexed by 1.

W2: If $L_1, L_2 \in \mathcal{W}$ and K_2 is a component of L_2 indexed by 0 or 2, then $L_1 \circ (L_2 - K_2) \circ u \in \mathcal{W}$.

W3: If $L_1, L_2 \in \mathcal{W}$ and K_1, K_2 are components of L_1, L_2 with indices 0 and 2 (resp.), then $(L_1 - K_1) \circ (L_2 - K_2) \circ u \in \mathcal{W}$.

W4: If $L_1, L_2 \in \mathcal{W}$ and K_1, K_2 are components of L_1, L_2 (resp.) each with index 0 or 2, then

$$((L_1, K_1)\#(L_2, K_2)) \cup m \in \mathcal{W},$$

where $K_1\#K_2$ shares the index of either K_1 or K_2 and m is a meridian of $K_1\#K_2$ indexed by 1.

W5: If $L \in \mathcal{W}$ and K is a component of L indexed by $i = 0$ or 2, then $L' \in \mathcal{W}$, where L' is obtained from L replacing a tubular neighborhood of K with a solid torus with three closed orbits, K_1, K_2, and K_3. K_1 is the core and so has the same knot type as K. K_2 and K_3 are parallel (p, q) cables of K_1. The index of K_2 is 1. The indices of K_1 and K_3 may be either 0 or 2, but at least one of them must be equal to the index of K.

W6: If $L \in \mathcal{W}$ and K is a component of L indexed by $i = 0$ or 2, then $L' \in \mathcal{W}$, where L' is obtained from L by changing the index of K to 1 and placing a $(2, q)$-cable of K in a tubular neighborhood of K, indexed by i.

W7: $\mathcal{W}$ is minimal. That is, $\mathcal{W} \subset \mathcal{W}'$ for any collection, $\mathcal{W}'$, satisfying W0-W6.

Remark A.1.12 The last condition, $W7$, means that $\mathcal{W}$ is generated from the indexed Hopf link in S^3 by applying *operations W1-W6*.

Theorem A.1.13 (Wada [184]) *Let $\mathcal{F}$ be the set of indexed links which can be realized as the collection of periodic orbits of a NMS on S^3, respecting index. Then $\mathcal{W} = \mathcal{F}$.*

Outline of proof: The argument for $\mathcal{W} \subset \mathcal{F}$ is straightforward though tedious. We must show that $\mathcal{F}$ obeys axioms $W0$ through $W6$. Example A.1.2 establishes $W0$. Example A.1.7 shows axiom $W6$ can be realized and Example A.1.9 can be generalized to show $\mathcal{F}$ obeys $W1$. The remaining axioms can be similarly shown to hold by explicit constructions.[3]

The proof of $\mathcal{F} \subset \mathcal{W}$ uses an induction strategy. Let $\mathcal{F}_r$ be the subcollection of $\mathcal{F}$ whose elements have at most r components of index 1. For $r = 0$, $\mathcal{F}_0$ contains just the Hopf link with indices 0 and 2. Thus, $\mathcal{F}_0 \subset \mathcal{W}$. Now suppose that for some $r \geq 1$, $\mathcal{F}_{r-1} \subset \mathcal{W}$. Let $L \in \mathcal{F}_r$. The corresponding flow has a round handle decomposition. By careful surgery, one removes a 1-RH from this flow and shows that two new flows on S^3 can be constructed from the remaining round handles. These flows have at least one fewer index 1 orbit and so are in $\mathcal{W}$. But the surgery is performed so that the process can be reversed via one of the moves $W1, \ldots, W6$. Hence, $\mathcal{F}_r$ is in $\mathcal{W}$ for all r. □

---

[3]The only construction which is very difficult is that of $W4$ — forming the connected sum. The summary article [35] contains a helpful diagram.

A.1.4   Extensions and applications

Remark A.1.14 Fomenko [49] has developed a general program for studying integrable Hamiltonian flows on three-manifolds which has fundamental connections to nonsingular Morse-Smale flows. Consider a symplectic four-manifold M with Hamiltonian H, a nondegenerate constant-energy three-manifold $Q = H^{-1}(c) \subset M$, and an additional integral F defined on a neighborhood of Q whose critical points in Q form nondegenerate submanifolds. Then, we say the Hamiltonian system defined by H is *Bott-integrable* on Q. This is a more general notion than that of (complete) integrability, in which every constant-energy submanifold is integrable.

For a Bott-integrable system on Q, there is a finite collection of critical submanifolds of F on Q which are periodic orbits: these form a link L_F in Q. The only other critical submanifolds present are singular tori. By the Liouville Theorem [6], the complement of the critical submanifolds of F in Q is foliated by tori. Any component of L_F is indexed with the index inherited from F. Knots of index zero or two (local minima/maxima of F) possess tubular neighborhoods foliated by tori except at the core. Knots of index one lie on one or two "bifurcation" tori, which correspond to inflection points for F.

Fomenko and Nguyen [50], using topological and dynamical methods, were the first to show that each periodic orbit of the Hamiltonian flow on Q with index zero or two must be a generalized iterated torus knot: that is, it is formed from the unknot by the operations of cabling and connected sum. Cassasays, Nunes, and Martínez Alfaro [35] revisit this work and point out that the Bott-integrable energy manifold Q must also support a NMS flow with cores of the RH decomposition related to the link L_F in a natural way. Thus, they conclude that the class of indexed links realizable as the set of stable periodic orbits for some H and F is generated by the axioms *W0, W4, W5, W6,* and *W7* of Definition A.1.11.

From these two works, it follows that *any* periodic orbit in the integrable Hamiltonian flow on Q must be a generalized iterated torus knot. See [35, 50] for definitions and further details.

Remark A.1.15 In [157], Saito extends Wada's theorem. Given any indexed link L and any 3-manifold M we cannot in general expect there to be a NMS flow on M, let alone one with nonwandering set L. However, Saito develops a canonical procedure for producing a new indexed link L', derived from any L, and a new manifold M' derived from M, such that there is a NMS flow on M' with nonwandering set L'. There are some minor restrictions on the initial link L and M must be orientable.

Remark A.1.16 Generalized iterated torus knots manifest themselves in other settings as well. Let ξ be a smooth plane field on S^3: that is, in the tangent space at each point there is a plane. Consider the class of vector fields which lie entirely within ξ. Such flows have characteristics of both two- and three-dimensional dynamics and arise in the study of *contact geometry*.

In [45], it is shown that [generic] singularities of a plane-field flow arise not in isolated points, but in embedded circles. Hence, the singularities of such a flow gives a link. Consider the class of flows with the simplest dynamics: gradient-like flows, for which the only recurrence is fixed points. Then the only types of links which may arise are the links described in Wada's Theorem.

A.2   Smale flows, abstract

In this section we review the work of Franks and others on *Smale flows*, especially nonsingular Smale flows on S^3. These results rely on the homology theory of filtrations associated to the flow. As this is outside the scope of this work, we will merely state results and outline applications. Thus, no use of homology will be made here. The interested reader should consult [53] as well as the references given there.

The theory outlined culminates in an abstract classification of Smale flows on S^3 using a device called the *Lyapunov graph*. By *abstract*, we mean that the embedding types of the basic sets are not determined, only which combinations of basic sets can be realized. The next section of this appendix addresses the question of how they may and may not fit together with respect to embedding.

Smale flows satisfy the same hyperbolicity and transversality conditions as Morse-Smale flows, but the basic sets may have infinitely many periodic orbits, while still being one-dimensional (or zero-dimensional if we allow for singularities). Recall from §1.2 the definition of a Smale flow:

Definition A.2.1 A flow ϕ_t on a manifold M is called a *Smale flow* if

- the chain recurrent set R of ϕ_t has a hyperbolic structure,

- the basic sets of R are zero- or one-dimensional, and

- the stable manifold of any orbit in R has transversal intersection with the unstable manifold of any other orbit of R.

Smale flows on compact manifolds are structurally stable under C^1 perturbations but are not dense in the space of C^1 flows. It is easy to see that for dim $M = 3$, each attracting and repelling basic set is either a closed orbit or fixed point. The admissible saddle sets, however, include suspensions of irreducible subshifts of finite type and can be nontrivial, i.e. they can have infinitely many closed orbits. Thus, while there are no strange attractors or repellors, complicated saddle sets may exist, which can be modeled by templates. Indeed, as we shall see, a suspension of the horseshoe, together with an attractor-repellor pair of periodic orbits, provides an important example of a nonsingular Smale flow.

Given a suspended subshift of finite type we can construct a Markov partition and a corresponding transition matrix A. We can encode additional information about the embedding of a basic set by modifying the transition matrix:

Definition A.2.2 Given a Markov partition for a cross section of a basic set with first return map ρ, assign an orientation to each partition element. If the partition is fine enough the function

$$O(x) = \left\{ \begin{array}{ll} +1 & \text{if } \rho \text{ is orientation preserving at } x, \\ -1 & \text{if } \rho \text{ is orientation reversing at } x, \end{array} \right.$$

is constant on each partition element. The *structure matrix* S is then defined by $S_{ij} = O(x)A_{ij}$, where x is any point in the i-th partition element. (This is slightly different then the structure matrix defined in §5.2.)

Example A.2.3 For a suspension of the full shift on two symbols modeled in a flow by the Lorenz template, $\begin{bmatrix} 1 & 1 \\ 1 & 1 \end{bmatrix}$ is the structure matrix. However, if the suspension of the full two-shift is modeled by the horseshoe template, then $\begin{bmatrix} 1 & 1 \\ -1 & -1 \end{bmatrix}$ is the corresponding structure matrix.

Later, we will define the linking matrix of a saddle set in a Smale flow that encodes how the orbits in the saddle set link the attracting and repelling orbits in the flow.

The suspension of any irreducible subshift of finite type can be realized as a basic set in a Smale flow on any manifold of dimension three [148] or greater [191]. The technique of [148] typically introduces many singularities. Franks [54] has observed that the realization result in [148] holds true for any structure matrix.

Theorem A.2.4 (Franks [54]) *Suppose S is an irreducible integer matrix. Then there exists a nonsingular Smale flow ϕ_t on some 3-manifold with basic set Λ whose structure matrix is S. It is possible to choose ϕ_t so that each basic set of ϕ_t, except for Λ, consists of a single closed orbit.*

Theorem A.2.5 (Franks [54]) *Suppose ϕ_t is a nonsingular Smale flow on S^3 with a basic set having an $n \times n$ structure matrix S. Then if $\det(I - S) \neq 0$, the group $\mathbb{Z}^n/(I - S)\mathbb{Z}^n$ must be cyclic.*

Example A.2.6 The matrix $S = \begin{bmatrix} 1 & 2 \\ 2 & 1 \end{bmatrix}$ cannot be realized as the structure matrix of a nonsingular Smale flow on S^3, since the quotient group $\mathbb{Z}^2/(I-S)\mathbb{Z}^2$ has presentation $\langle x, y : 2x = 2y = 0 \rangle$, which is isomorphic to $\mathbb{Z}_2 \oplus \mathbb{Z}_2$.

Suppose there is a single attracting closed orbit γ_a, and a single repelling closed orbit γ_r, with all other basic sets saddles. Then we may compute the absolute value of the linking number of γ_a and γ_r as follows. Let $\Lambda_1, \dots \Lambda_n$ denote the saddle sets and let $S_1 \dots S_n$ denote the respective structure matrices. It is shown in [51] that

$$|\ell k\,(\gamma_a, \gamma_r)| = \prod_{i=1}^{n} |\det(I - S_i)|,$$

where the product is taken to be one if $n = 0$. We remark that $|\det(I - S_i)|$ is the order of the group $\mathbb{Z}^m/(I - S_i)\mathbb{Z}^m$ where m is the size of S_i.

Example A.2.7 Given a flow as above with a single saddle set having structure matrix $\begin{bmatrix} -1 & -1 \\ -1 & -1 \end{bmatrix}$, γ_a and γ_r have linking number three. Figure A.7 depicts a realization of this example. The figure shows an isolating neighborhood for each of the three basic sets. For γ_a and γ_r, these are the solid tori V_a and V_r respectively. Call the saddle set Λ and its isolating neighborhood N. Now N is isotopic to the unit normal bundle of a template $\mathcal{T}$. The template $\mathcal{T}$ is shown in Figure A.8, where we see how to isotope it to look more like the templates presented in earlier chapters. The exit set of N is isotopic to the unit normal bundle over $\partial\mathcal{T}$ and is attached to ∂V_a. We can now see how to attach ∂V_r to $\partial(V_a \cup N)$ and form S^3.

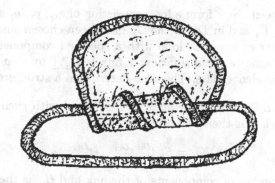

Figure A.7: A Smale flow with $\ell k\,(\gamma_a, \gamma_r) = 3$.

If we know how the saddle sets "link" a collection L of attracting and repelling closed orbits we can say more: we can compute a polynomial invariant of the link L. This invariant is none other than the Alexander polynomial, a standard invariant of classical knot theory [154, 33].

The manner in which a saddle set "links" a collection of closed orbits is described by modifying the structure matrix S to form a *linking matrix* K. Consider a cross section of the saddle set that is homeomorphic to a subshift of finite type $\sigma : \Sigma_A \to \Sigma_A$, by a homeomorphism h. We define Cantor sets $\{C_i\}_{i=1}^n$ by $C_i = h(\{\mathbf{a} \in \Sigma_A | a_0 = x_i\})$. As in Lemma 2.2.5, we can extend the $\{C_i\}_{i=1}^n$ to two-dimensional disks $\{D_i\}_{i=1}^n$ which are transverse to the ambient flow such that (a) $C_i = D_i \cap S$, (b) $\partial D_i \cap R = \emptyset$, and (c) $D_i \cap L = \emptyset$, for $i = 1, ..., n$.

Next we pick a base point b in $S^3 - L$ and paths p_i from b to D_i, also in $S^3 - L$. Let γ_{ij} be a segment of the flow going from C_i to C_j without meeting any

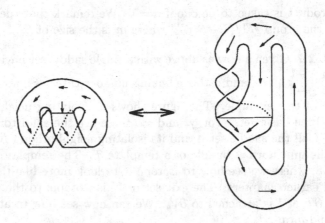

Figure A.8: A template for the flow in Figure A.7.

of the C_k in between. Now form a loop consisting of γ_{ij}, p_i, p_j and, if needed, a short segment in D_i and in D_j. If the C_k have been chosen small enough, then the linking number of any such loop with a specified component of L depends only on i and j. One can find sufficiently fine $\{C_i\}_{i=1}^n$ by changing the matrix A in its shift equivalence class. This also determines a structure matrix S.

Definition A.2.8 The *linking matrix* K associated with such a choice of the C_k for a given link L is then defined to be

$$K_{ij} = S_{ij} t_1^{\ell k_1} t_2^{\ell k_2} \cdots t_\mu^{\ell k_\mu},$$

where μ is the number of components of the link and ℓk_p is the linking number of the loops formed from segments connecting C_i to C_j and the pth component of L.

Theorem A.2.9 (Franks [52]) *Suppose that ϕ_t is a nonsingular Smale flow on S^3, L is a μ-component link of closed orbits oriented by the flow, each an attractor or repellor, and that $\{K_i\}_{i=1}^n$ are linking matrices of the saddle sets with respect to L. Let m_{ij} denote the linking number of the ith component of L with the jth component of the set of attractors and repellors not in L. If $\mu = 1$, i.e., L is a knot, then*

$$\Delta_L(t) = \frac{(1-t)\prod_i \det(I - K_i)}{\prod_k (1 - t^{m_{1k}})},$$

is an isotopy invariant of the oriented knot, up to multiples of $\pm t^{\pm 1}$. This invariant is precisely the Alexander polynomial of the knot [154, 33]. If, if $\mu > 1$,

$$\Delta_L(t_1, ..., t_\mu) = \frac{\prod_i \det(I - K_i)}{\prod_k (1 - t^{m_{1k}} \cdots t^{m_{\mu k}})},$$

is an isotopy invariant of the oriented link, up to multiples of $\pm t_j^{\pm 1}$. Again, this invariant is the Alexander polynomial of the link L.

Example A.2.10 Figure A.9 shows a Smale flow with three basic sets. The attractor γ_a is a trefoil knot. The saddle set can clearly be modeled by a Lorenz template. Using the obvious two-element Markov partition for the Lorenz template, we find that a linking matrix for the saddle set with respect to the one-component link γ_a is $\begin{bmatrix} t & t \\ 1/t & 1/t \end{bmatrix}$. Thus, the Alexander polynomial of γ_a is $-t^{-1} + 1 - t$. Any isolated closed orbit in a Smale flow which has polynomial different from this, up to multiples of t, cannot be isotopic to the trefoil.

Finally, in [56] we have an *abstract* classification of nonsingular Smale flows on S^3. The major new tool is the *Lyapunov graph*. Given a Smale flow on a manifold there exists a smooth function from the manifold to the reals which is non-increasing with respect to the flow (time) parameter [53, pages 1 and 2]. Thus, each basic set is mapped to a point. This is called a *Lyapunov function*. The *Lyapunov graph* is defined by identifying connected components of the inverse images of points in the real line. Each vertex of the graph is a point whose connected component contains a basic set. Vertices is labeled by the corresponding basic sets and edges are oriented by the flow direction.

Suppose Γ is an abstract Lyapunov graph whose sinks and sources are each labeled with a single attracting or repelling periodic orbit and suppose each remaining vertex is labeled with the suspension of a subshift of finite type. Then Γ is associated with a nonsingular Smale flow on S^3 if and only if the following are satisfied: (1) The graph Γ is a tree with one edge attached to each source and each sink vertex. (2) If v is a saddle vertex whose basic set has transition matrix A and with e_v^+ entering edges and e_v^- exiting edges then

$$e_v^+ \leq Z_A + 1$$
$$e_v^- \leq Z_A + 1$$
$$Z_A + 1 \leq e_v^+ + e_v^-.$$

Here, Z_A is a the *Zeeman number* defined by $\dim \ker((I - A_2) : \mathbb{Z}_2^n \to \mathbb{Z}_2^n)$, where A_2 is the mod 2 reduction of A, $\mathbb{Z}_2$ is the integers mod 2, and n is the size of A.

An abstract classification theorem for Smale flows in S^3 with singularities has been obtained by de Rezende [40].

A.3 Smale flows, embedded

The contrast between Smale and Morse-Smale flows reveals itself not only in the saddle sets, but also in the embedding of the isolated periodic orbits. For a nonsingular Smale flow on S^3, *any* link can be the attractor, in contrast to the restricted class described in Wada's Theorem A.1.13.

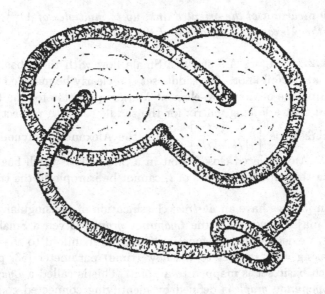

Figure A.9: A Smale flow with an attracting trefoil, γ_a, a Lorenz saddle set, and an unknotted repellor, γ_r.

Theorem A.3.1 (Franks [52]) *If L is any smooth link in S^3 then there exists a nonsingular Smale flow ϕ_t on S^3 such that L is the set of attractors and ϕ_t has a single unknotted repellor.*

Outline of Proof: Consider a disk D^2 with n distinguished points placed along a line within D^2. There exists a Smale diffeomorphism from D^2 into itself which fixes this set of n points as attractors, permutes two adjacent points, and fixes the $n - 2$ remaining points individually. Of course, several saddle points must also exist, to separate the domains of attraction. The suspension of this diffeomorphism can be embedded so that the trajectories on the n distinguished attracting points trace out the closure of a standard generator σ_i of the braid group B_n (*cf.* §1.1): see Figure A.10. Then, the suspension flow is a Smale flow, in-flowing on $\partial D^2 \times S^1$.

By suspending the composition of several such Smale diffeomorphisms, one may form a nonsingular Smale flow on a solid torus having any braid as an attractor. Some care is needed to make sure the vector field is smooth. Since any link can be braided (Theorem 1.1.13), adding a single repellor in the complementary solid torus yields the desired result.   □

Remark A.3.2 Notice that, in this construction, the repellor links the attracting link n times. That is, the sum of the linking numbers of the repellor over all the components of the attractor is n. Theorem A.3.1 may be refined to show that the repellor need not link the attractor at all.

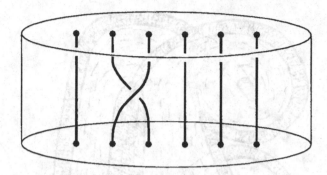

Figure A.10: The suspension of a disk map in which the saddle points trace out a braid.

As a final variation on this theme we prove the following result, which is a bit weaker, but has an interesting proof:

Theorem A.3.3 *If L is any smooth link in S^3, then there exists a nonsingular Smale flow with one saddle set such that L is a subcollection of the set of attractors, and such that there is a unique repellor which, together with one other attractor, forms a Hopf link separable from L.*

Proof: Figure A.11 shows a Smale flow whose saddle set can be modeled with the template V from §3.2. The attracting and repelling orbits form a Hopf link which can separated from the saddle set by a 2-sphere.

Recall the DA move for templates, related to the DA procedure of §2.2.2, and used on the horseshoe template in §4.2.1: this involves splitting a template T along a periodic orbit K to obtain a new template $DA_K(T)$ with K as an attractor. Figure A.12 shows this process for an orbit on V. Now, if T is a model of a saddle set in some Smale flow, we may form a new Smale flow, replacing T with a saddle set modeled by $DA_K(T)$ and a new attracting orbit with knot type K, linking each orbit in $DA_K(T)$ just as K did. By looking at the action on branch line charts, it is clear that this splitting on a connected template yields a connected template; all other basic set are unchanged. In Figure A.13, we show the result of this construction on the Smale flow of Figure A.11 using the orbit depicted in Figure A.12.

By Theorem 3.2.8, the link L is in V as a collection of closed orbits $K_1, \ldots, K_n$. We apply the DA process above to $K_1, \ldots, K_n$ successively to produce the desired flow.   □

Remark A.3.4 We now have a method for creating new Smale flows from old ones that at least suggests a bifurcation process, much as in Examples A.1.7 and A.1.9 of §A.1.

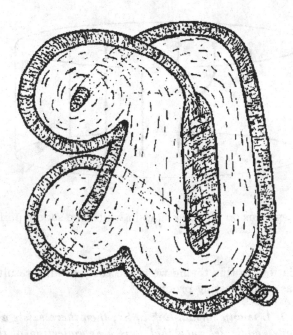

Figure A.11: A universal template $\mathcal{V}$ in a Smale flow.

A.3.1   Lorenz templates

We now consider the problem of realizing Smale flows from another viewpoint. Suppose we have a nonsingular Smale flow of S^3 with three basic sets, a unique attracting closed orbit, a unique repelling closed orbit, and a unique saddle set modeled topologically by a Lorenz template. That is, there exists a neighborhood of the saddle set foliated by local stable manifolds, such that when the leaves of the stable manifolds are identified, we get an embedding of the Lorenz template $\mathcal{L}(0,0)$. Let N_a, N_r and $N_{\mathcal{L}}$ be isolating tubular neighborhoods of the attractor, the repellor and the saddle set respectively. We ask: what are all the possible configurations of such a system? We want to classify the embeddings of N_a, N_r and $N_{\mathcal{L}}$ up to ambient isotopy, mirror images and flow reversal. To date, it is possible only to give a partial answer.

We start by showing in Figure A.14 an isolating neighborhood, $N_{\mathcal{L}}$, of the Lorenz saddle set glued to a 3-ball along its exit set. Topologically, the union it just a 3-ball itself. Thus, we may build a flow consisting of an attracting fixed point in the original 3-ball, the Lorenz saddle set, and a repelling fixed point in S^3 minus the Lorenz union 3-ball.

Figure A.14 also shows two ways one might attach *handles* to the 3-ball so as to turn it into a solid torus. Suppose we attach the handle to to the small disks marked C and C' in the manner shown. Call the resulting solid torus N_a'. If we take $N_{\mathcal{L}} \cup N_a'$ the result is still a solid torus, and the complement in S^3 is

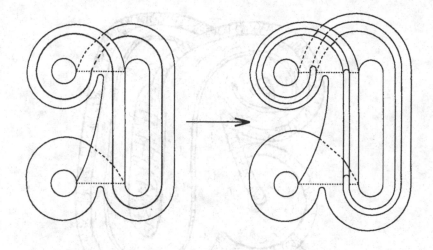

Figure A.12: The DA move on a closed orbit in $\mathcal{V}$.

just another solid torus, N_r'. We may now build a Smale flow with an attractor in N_a', a repellor in N_r' and a Lorenz saddle set in $N_{\mathcal{L}}$.

The exit set of $N_{\mathcal{L}}$ contains two annuli which are labeled X and Y in the figure. Call the cores of X and Y, x and y respectively. The reader should check that x and y each bound disks in $\partial N_a'$.

Upon further inspection the reader should be able to see that x and y can be made parallel. To be more precise, y and x together form the boundary of an annulus in $\partial N_a'$.

Now, instead on attaching a handle at C and C', attach one to B and B' as shown again in Figure A.14. This time call the solid torus obtained N_a''. As before $N_{\mathcal{L}} \cup N_a''$ is a solid torus with solid torus complement in S^3. Thus we have a Smale flow. Is it the same as the previous example?

To see that these flows differ, consider again the loops x and y. They are still both inessential, that is they both bound disks in $\partial N_a''$. But they are no longer concentric. This can be seen from careful study of the figure.

These two examples are the only Smale flows with the three basic sets we specified with both the loops x and y inessential in the boundary of the tubular neighborhood of the attracting orbit. We shall not prove this fact here, though the argument is quite standard.

In order to complete our task we have to consider two more cases, x and y both essential and one essential and the other not. An example of the latter can be obtained by attaching a handle to the disks on the 3-ball labeled A and B in Figure A.14. It can be shown that if y is essential and x is not, then the annulus Y can have any number of full twists if y is unknotted. If y is knotted, it must be a torus knot, and the amount of twist is fixed by the knot type of y. In all cases X is untwisted and the attractor-repellor pair forms a Hopf link. In Figure A.15 we show the y loop is a (2,1) curve on A. Detailed proofs of these

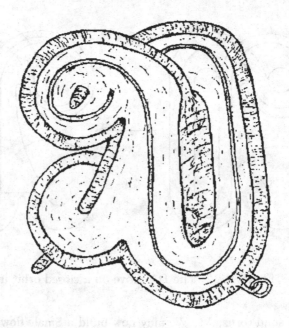

Figure A.13: The template $DA_K(\mathcal{V})$ in a Smale flow.

claims can be found in [172].

For an example of both x and y essential, connect a handle to the disks A and C, so that the complement in S^3 is an unknotted solid torus. This was shown above in Figure A.9. The attractor is a trefoil knot. It is shown in [172] that, up to mirror images and flow reversal, this is the only case for x and y both essential.

It is unlikely that there will ever be as complete an understanding of Smale flows, even nonsingular ones in S^3, as Wada and others have provided for non-singular Morse-Smale flows in S^3. However, we hope that the tools sketched here and currently under development will enable researchers to analyze those Smale flows in 3-manifolds that are of special interest to them.

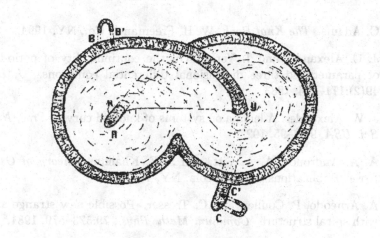

Figure A.14: A neighborhood of the Lorenz saddle set is glued to a 3-ball.

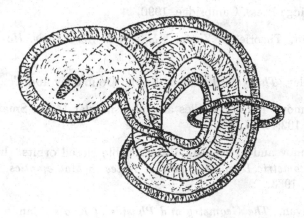

Figure A.15: The y loop is a (2,1) cable.

Bibliography

[1] C. Adams. *The Knot Book*. W. H. Freeman & Co., NY, 1994.

[2] J. C. Alexander and J. Yorke. On the continuability of periodic orbits of parametrized three dimensional differential equations. *J. Diff. Eq.*, 49(2):171–184, 1983.

[3] J. W. Alexander. A lemma on systems of knotted curves. *Proc. Nat. Acad. Sci. USA*, 9:93–95, 1923.

[4] A. A. Andronov, A. A. Vitt, and S. E. Khaikin. *Theory of Oscillators*. Dover Publications Inc., Mineola, NY, 1987.

[5] A. Arnéodo, P. Coullet, and C. Tresser. Possible new strange attractors with spiral structure. *Commun. Math. Phys.*, 79:573–579, 1981.

[6] V. I. Arnold. *Geometrical Methods in the Theory of Ordinary Differential Equations*. Springer-Verlag, Berlin, Heidelberg, New York, 1982.

[7] V. I. Arnold. The asymptotic Hopf invariant and its applications. *Sel. Math. Sov.*, 5:327–345, 1986.

[8] V. I. Arnold and A. Avez. *Ergodic Problems of Classical Mechanics*. Benjamin, New York, 1968.

[9] D. Arrowsmith and C. Place. *An Introduction to Dynamical Systems*. Cambridge Press, Cambridge, 1990.

[10] E. Artin. Theorie der Zopfe. *Abh. Math. Sem. Univ. Hamburg*, 4:47–72, 1925.

[11] C. Ashley. *The Ashley Book of Knots*. Doubleday and Co., NY, 1944.

[12] D. Asimov. Round handles and non-singular Morse-Smale flows. *Ann. Math.*, 102:41–54, 1975.

[13] D. Asimov and J. Franks. Unremovable closed orbits. In J. Palis, editor, *Geometric Dynamics, Lecture Notes in Mathematics 1007*. Springer-Verlag, 1983.

[14] M. Atiyah. *The Geometry and Physics of Knots*. Cambridge University Press, Cambridge, 1990.

[15] H. Bell and K. Meyer. Limit periodic functions, adding machines, and solenoids. Preprint, Apr. 1995.

[16] V. Belykh and L. Chua. New type of strange attractor from a geometric model of Chua's circuit. *Int. J. Bif. Chaos*, 2(3):697–704, 1992.

[17] D. Bennequin. Entrelacements et équations de Pfaff. *Asterisque*, 107-108:87–161, 1983.

[18] G. D. Birkhoff. *Dynamical Systems*. AMS, Providence, RI, 1927.

[19] J. Birman. *Braids, Links, and Mapping Class Groups*. Princeton University Press, Princeton, N.J., 1974.

[20] J. Birman. Recent developments in braid and link theory. *Math. Intelligencer*, 13(1):52–60, 1991.

[21] J. Birman. New points of view in knot theory. *Bull. Am. Math. Soc.*, 28(2):253–285, 1994.

[22] J. Birman and X. S. Lin. Knot polynomials and Vassiliev invariants. to appear in *Invent. Math.*, preprint, 1995.

[23] J. Birman and R. Williams. Knotted periodic orbits in dynamical systems–I : Lorenz's equations. *Topology*, 22(1):47–82, 1983.

[24] J. Birman and R. Williams. Knotted periodic orbits in dynamical systems–II : knot holders for fibered knots. *Cont. Math.*, 20:1–60, 1983.

[25] R. Bowen. One-dimensional hyperbolic sets for flows. *J. Diff. Eq.*, 12:173–179, 1972.

[26] R. Bowen. On Axiom A diffeomorphisms. In *Regional Conference Series in Mathematics 35*, pages 1–45. National Science Foundation, American Mathematical Society, 1978.

[27] R. Bowen and J. Franks. Homology of zero dimensional basic sets. *Ann. Math.*, 106:73–92, 1977.

[28] R. Bowen and P. Walters. Expansive one-parameter flows. *J. Diff. Eq.*, 12:180–193, 1972.

[29] P. Boyland. Braid-types of periodic orbits for surface automorphisms. In *Notes on Dynamics of Surface Homeomorphisms*. Warwick University, UK, Lecture notes, Maths. Inst., 1989.

[30] P. Boyland. Topological methods in surface dynamics. *Topology and its Applications*, 58:223–298, 1994.

[31] P. Boyland and T. Hall. *Dynamics of Surface Homeomorphisms*. World Scientific, Singapore, to appear, preprint.

[32] M. Boyle. Symbolic dynamics and matrices. In R.A. Bruald, S. Freedland, and V. Klee, editors, *Combinatotial and graph-theoretical problems in linear algebra*, volume 50 of *IMA Math. Appl.*, pages 1–38. Springer-Verlag, New York, 1993.

[33] G. Burde and H. Zieschang. *Knots*. De Gruyter, Berlin, 1985.

[34] J. Carr. *Applications of Centre Manifold Theory*. Springer-Verlag, Berlin, Heidelberg, New York, 1981.

[35] J. Casasayas, J. Martínez Alfaro, and A. Nunes. Knots and links in integrable Hamiltonian systems. Preprint, Warwick 48/1992, Sept. 1992.

[36] A. Champneys. Homoclinic tangencies in the dynamics of articulated pipes conveying fluid. *Physica D*, 62:347–359, 1993.

[37] J. Christy. Branched surfaces and attractors I: dynamic branched surfaces. *Trans. Am. Math. Soc.*, 336(2):759–784, 1993.

[38] L. Chua, M. Komuro, and T. Matsumoto. The double scroll family. *IEEE Trans. on Circuits and Systems*, 33:1073–1118, 1986.

[39] P. Collet and J. P. Eckmann. *Iterated Maps on the Interval as Dynamical Systems*. Birkhauser, Boston, 1980.

[40] K. de Rezende. Smale flows on the three-sphere. *Trans. Am. Math. Soc.*, 303:283–310, 1987.

[41] R. Devaney. *An Introduction to Chaotic Dynamical Systems*. Benjamin Cummings, Menlo Park, CA, 1985.

[42] R. Devaney and Z. Nitecki. Shift automorphisms in the Hénon mapping. *Commun. Math. Phys.*, 67:137–148, 1979.

[43] F. Diacu and P. Holmes. *Celestial Encounters: The Origins of Chaos and Stability*. Princeton University Press, Princeton, NJ, 1996.

[44] F. Dumortier, H. Kokubu, and H. Oka. A degenerate singularity generating geometric Lorenz attractors. *Ergod. Th. & Dynam. Sys.*, 15:833–856, 1995.

[45] J. Etnyre and R. Ghrist. The dynamics of legendrian flows. Unpublished notes, Oct. 1996.

[46] A. Fathi, F. Laudenbach, and V. Poenaru et al. Travaux de Thurston sur les surfaces. *Astérique*, 66-67:1–284, 1979.

[47] M. J. Feigenbaum. Quantitative universality for a class of nonlinear transformations. *J. Stat. Phys.*, 19:25–52, 1978.

[48] W. Floyd and U. Oertel. Incompressible surfaces via branched surfaces. *Topology*, 23:117–125, 1984.

[49] A. T. Fomenko. *Advances in Soviet Mathematics, Vol. 6*. American Mathematical Society, Providence, RI, 1991.

[50] A. T. Fomenko and T. Z. Nguyen. Topological classification of integrable nondegenerate Hamiltonians on isoenergy three-dimensional spheres. In A. T. Fomenko, editor, *Advances in Soviet Mathematics, Vol. 6*, pages 267–296. American Mathematical Society, 1991.

[51] J. Franks. Non-singular flows on S^3 with hyperbolic chain-recurrent set. *Rocky Mountain J. Math.*, 7(3):539–546, 1977.

[52] J. Franks. Knots, links, and symbolic dynamics. *Ann. of Math.*, 113:529–552, 1981.

[53] J. Franks. *Homology and Dynamical Systems*, volume CBMS 49. American Mathematical Society, Providence, Rhode Island, 1982.

[54] J. Franks. Symbolic dynamics in flows on three-manifolds. *Trans. Am. Math. Soc.*, 279:231–236, 1983.

[55] J. Franks. Flow equivalence of shifts of finite type. *Ergodic Theory and Dynamical Systems*, 4(1):53–66, 1984.

[56] J. Franks. Nonsingular Smale flows on S^3. *Topology*, 24:265–282, 1985.

[57] J. Franks and C. Robinson. A quasi-Anosov diffeomorphism which is not Anosov. *Trans. Am. Math. Soc.*, 223:267–278, 1976.

[58] J. Franks and R. F. Williams. Entropy and knots. *Trans. Am. Math. Soc.*, 291(1):241–253, 1985.

[59] P. Freyd, D. Yetter, J. Hoste, W. B. R. Lickorish, K. Millet, and A. Ocneanu. A new polynomial invariant of knots and links. *Bull. Am. Math. Soc.*, 12(2):239–246, 1985.

[60] G. Frobenius. Über Matrizen aus nicht negativen Elementen. *S.B.-Deutsch. Akad. Wiss. Berlin Math.-Nat. Kl.*, pages 456–477, 1912.

[61] D. Gabai. Foliations and the topology of 3-manifolds III. *J. Diff. Geom.*, 26:479–536, 1987.

[62] J.-M. Gambaudo and E. Ghys. Entrlacements asymptotiques. To appear in *Topology*, preprint, August 1996.

[63] J.-M. Gambaudo, P. Glendinning, and C. Tresser. The gluing bifurcation I: symbolic dynamics of the closed curves. *Nonlinearity*, 1:203–214, 1988.

[64] J.-M. Gambaudo, D. Sullivan, and C. Tresser. Infinite cascades of braids and smooth dynamical systems. *Topology*, 33(1):85–94, 1994.

[65] J.-M. Gambaudo and C. Tresser. Dynamique régulière ou chaotique: applications du cercle ou de l'intervalle ayant une discontinuité. *C. R. Acad. Sc. Paris*, 300:311–313, 1985.

[66] P. Gaspard, R. Kapral, and G. Nicolis. Bifurcation phenomena near homoclinic systems: a two-parameter family. *J. Stat. Phys.*, 35:697–727, 1984.

[67] K. F. Gauss. Zur mathematischen Theorie der electrodynamischen Wirkungen. *Werke Konigl. Gesell. Wiss. Gottingen*, 5:605, 1877.

[68] R. Ghrist. Accumulations of infinite links in flows. Preprint, August 1996.

[69] R. Ghrist. Branched two-manifolds supporting all links. *Topology*, 36(2):423–448, 1997.

[70] R. Ghrist and P. Holmes. Knots and orbit genealogies in three dimensional flows. In *Bifurcations and Periodic Orbits of Vector Fields*, pages 185–239. NATO ASI series C volume 408, Kluwer Academic Press, 1993.

[71] R. Ghrist and P. Holmes. An ODE whose solutions contain all knots and links. *Intl. J. Bifurcation and Chaos*, 6(5):779–800, 1996.

[72] P. Glendinning. Global bifurcation in flows. In T. Bedford and J. Swift, editors, *New Directions in Dynamical Systems*, pages 120–149. London Math. Society, Cambridge University Press, 1988.

[73] P. Glendinning, J. Los, and C. Tresser. Renormalisation between classes of maps. *Phys. Lett. A*, 145:109–112, 1990.

[74] P. Glendinning and C. Sparrow. Local and global behavior near homoclinic orbits. *J. Stat. Phys.*, 35:645–698, 1984.

[75] P. Glendinning and C. Sparrow. T-points: a codimension two heteroclinic bifurcation. *J. Stat. Phys.*, 43:479–488, 1986.

[76] J. Guckenheimer and P. Holmes. *Nonlinear Oscillations, Dynamical Systems, and Bifurcations of Vector Fields*. Springer-Verlag, New York, 1983.

[77] J. Guckenheimer and R. Williams. Structural stability of Lorenz attractors. *Inst. Hautes Études Sci. Publ. Math.*, 50:59–72, 1979.

[78] C. Gutierrez. Structural stability for flows on the torus with a cross-cap. *Trans. Am. Math. Soc.*, 241:311–320, 1978.

[79] C. Gutierrez. Knots and minimal flows on S^3. *Topology*, 241:679–698, 1995.

[80] P. Hadamard. Les surfaces a cobures opposées et leur lignes géodesiques. *J. Math. Pures Appl.*, 4:27–73, 1898.

[81] V. Hansen. *Braids and Coverings: Selected Topics*. Cambridge University Press, Cambridge, 1991.

[82] A. Hatcher. On the boundary curves of incompressible surfaces. *Pac. J. Math.*, 99(2):373–377, 1982.

[83] M. Hénon. A two dimensional map with a strange attractor. *Commun. Math. Phys.*, 50:69–77, 1976.

[84] M. Hirsch, C. Pugh, and M. Shub. *Invariant Manifolds*, volume 583 of *Springer Lecture Notes in Mathematics*. Springer-Verlag, Berlin, Heidelberg, New York, 1970.

[85] K. Hockett and P. Holmes. Josephson's junction, annulus maps, Birkhoff attractors, horseshoes and rotation sets. *Ergod. Th. and Dynam. Sys.*, 6:205–239, 1986.

[86] P. Holmes. A strange family of three-dimensional vector fields near a degenerate singularity. *J. Diff. Eq.*, 37(3):382–403, 1980.

[87] P. Holmes. Bifurcation sequences in the horseshoe map: infinitely many routes to chaos. *Phys. Lett. A*, 104:299–302, 1984.

[88] P. Holmes. Knotted periodic orbits in suspensions of Smale's horseshoe: period mutiplying and cabled knots. *Physica D*, 21:7–41, 1986.

[89] P. Holmes. Knotted periodic orbits in suspensions of annulus maps. *Proc. Roy. London Soc. A*, 411:351–378, 1987.

[90] P. Holmes. Knotted periodic orbits in suspensions of Smale's horseshoe: extended families and bifurcation sequences. *Physica D*, 40:42–64, 1989.

[91] P. Holmes and R. Ghrist. Knotting within the gluing bifurcaion. In J. M. T. Thompson and S. R. Bishop, editors, *IUTAM Symposium on Nonlinearity and Chaos in Engineering Dynamics*, pages 299–315. John Wiley, Chichester, UK, 1994.

[92] P. Holmes and D. C. Whitley. On the attracting set for Duffing's equation. *Physica D*, 7:111–123, 1983.

[93] P. Holmes and R. F. Williams. Knotted periodic orbits in suspensions of Smale's horseshoe: torus knots and bifurcation sequences. *Archive for Rational Mech. and Anal.*, 90(2):115 –193, 1985.

[94] V. F. R. Jones. A polynomial invariant for knots via von Neumann algebras. *Bull. Am. Math. Soc.*, 12(1):103–111, 1985.

[95] L. Jonker and D. A. Rand. Bifurcations in one dimension I: the nonwandering set. *Invent. Math.*, 62:347–365, 1981.

[96] L. Jonker and D. A. Rand. Bifurcations in one dimension II: a versal model for bifurcations. *Invent. Math.*, 63:1–15, 1981.

[97] A. Katok. Lyapunov exponents, entropy, and periodic orbits for diffeomorphisms. *Publ. Inst. Hautes Études Sci.*, 51:137–173, 1980.

[98] L. Kauffman. An invariant of regular isotopy. *Trans. Am. Math. Soc.*, 318:417–471, 1990.

[99] L. Kauffman. *Knots and Physics*. World Scientific, Singapore, 1991.

[100] L. Kauffman, M. Saito, and M. Sullivan. Quantum invariants for templates. Preprint, 1996.

[101] L. H. Kauffman. *On Knots*. Princeton University Press, Princeton, NJ, 1987.

[102] S. Kennedy. *A Lorenz-like Strange Attractor*. PhD thesis, Northwestern University, 1988.

[103] S. Kennedy, M. Staford, and R. Williams. A new Caylay-Hamilton theorem. In *Global analysis in modern mathematics*, pages 247–251, Houston, TX, 1993. Publish or Perish.

[104] P. Kent and J. Elgin. Noose bifurcation of periodic orbits. *Nonlinearity*, 4:1045–1061, 1991.

[105] A. Khibnik, D. Roose, and L. Chua. On periodic orbits and homoclinic bifurcations in Chua's circuit with a smooth nonlinearity. *Int. J. Bif. and Chaos*, 3(2):363–384, 1993.

[106] Lj. Kocarev, D. Dimovski, Z. Tasev, and L. Chua. Topological description of a chaotic attractor with spiral structure. *Phys. Lett. A*, 190:399–402, 1994.

[107] K. Kuperberg. A smooth counterexample to the Seifert conjecture. *Ann. of Math.*, 140:723–732, 1994.

[108] C. Letellier, P. Dutertre, and G. Gouesbet. Characterization of the Lorenz system, taking into account the equivariance of the vector field. *Phys. Rev. E*, 49(4):3492–3495, 1994.

[109] C. Letellier, P. Dutertre, and B. Maheu. Unstable periodic orbits and templates of the Rössler system: toward a systematic topological characterization. *Chaos*, 5:271–282, 1995.

[110] C. Letellier, P. Dutertre, J. Reizner, and G. Gouesbet. Evolution of a multimodal map induced by an equivariant vector field. *J. Phys. A: Math. Gen.*, 29:5359–5373, 1996.

[111] C. Letellier and G. Gouesbet. Topological characterization of a system with high order symmetries. *Phys. Rev. E*, 52(5):4754–4761, 1995.

[112] C. Letellier and G. Gouesbet. Topological characterization of reconstructed attractors modding out symmetries. *J. Phys. II France*, 6:1615–1638, 1996.

[113] C. Letellier, G. Gouesbet, F. Soufi, J.R. Buchler, and Z. Kolláth. Chaos in variable stars: topological analysis of W Vir model pulsations. *Chaos*, 6:466–476, 1996.

[114] E. Lorenz. Deterministic non-periodic flow. *J. Atmospheric Sci.*, 20:130–141, 1963.

[115] J. Los. Knots, braid index, and dynamical type. *Topology*, 33(2):257–270, 1994.

[116] L. Markus and K. Meyer. Periodic orbits and solenoids in generic Hamiltonian systems. *Amer. J. Math.*, 102:25–92, 1980.

[117] W. Massey. *Algebraic Topology, an Introduction*. Springer-Verlag, Berlin, Heidelberg, New York, 1967.

[118] F. McRobie and J. M. T. Thomson. Braids and knots in driven oscillators. *Int. J. Bifurcation and Chaos*, 3:1343–1361, 1993.

[119] F. McRobie and J. M. T. Thomson. Knot types and bifurcation sequences of homoclinic and transient orbits of a single-degree-of-freedom driven oscillator. Submitted to *Dynamics and Stability of Systems*, Preprint, April 1993.

[120] V. K. Melnikov. On the stability of the center for time-periodic perturbation. *Trans. Moscow Math. Soc.*, 12:1–57, 1963.

[121] N. Merener and G. Mindlin. From time series to physical models: the case of a pulsating star. Preprint, Departimento de Física, FCEN, Buenos Aires, Argentina, 1996.

[122] K. Meyer. Generic bifurcation of periodic points. *Trans. Am. Math. Soc.*, 149:95–107, 1970.

[123] K. Meyer. Generic stability properties of periodic points. *Trans. Am. Math. Soc.*, 154:273–277, 1971.

[124] J. Milnor. *Topology from the Differentiable Viewpoint*. University Press of Virginia, 1969.

[125] J. Milnor and W. Thurston. On iterated maps of the interval I and II. Unpublished notes, Princeton University, 1977.

[126] G. Mindlin and H. Solari. Topologically inequivalent embeddings. *Phys. Rev. E*, 52(2):1497–1501, 1995.

[127] G. Mindlin and H. Solari. Tori and Klein bottles in 4 dimensional chaotic flows. Preprint, Departimento de Fisica, FCEN, Buenos Aires, Argentina, 1996.

[128] G. Mindlin, H. Solari, M. Natiello, R. Gilmore, and X. Hou. Topological analysis of chaotic time series data from the Belousov-Zhabotinskii reaction. *J. Nonlinear Sci.*, 1:147–173, 1991.

[129] H. Moffat. Magnetostatic equilibria and analogous Euler flows of arbitrarily complex topology: part I. *J. Fluid Mech.*, 159:359–378, 1985.

[130] C. Moore. Braids in classical dynamics. *Phys. Rev. Lett.*, 70 (24):3675–3679, 1993.

[131] L. Mora and M. Viana. Abundance of strange attractors. *Acta Math.*, 171:1–71, 1993.

[132] J. Morgan. Nonsingular Morse-Smale flows on 3-dimensional manifolds. *Topology*, 18:41–54, 1978.

[133] M. Morse. Representations of geodesics. *Amer. J. Math.*, 43:33–51, 1921.

[134] M. Morse and G. Hedlund. Symbolic dynamics. *Amer. J. Math.*, 60:815–866, 1938.

[135] J. Moser. *Stable and Random Motions in Dynamical Systems*. Princeton University Press, Princeton, NJ, 1973.

[136] M. Muldoon, R. MacKay, J. Huke, and D. Broomhead. Topology from a time series. *Physica D*, 65:1–16, 1993.

[137] A. Nayfeh and D. Mook. *Nonlinear Oscillations*. John Wiley and Sons Inc., New York, 1979.

[138] J. Nielsen. Untersuchungen zur Topologie der gerschlossenen zweiseitegen Flächen I. *Acta. Math.*, 50:189–358, 1927.

[139] J. Palis and W. De Melo. *Geometric theory of dynamical systems: an introduction.* Springer-Verlag, Berlin, Heidelberg, New York, 1982.

[140] J. Palis and F. Takens. *Hyperbolicity and Sensitive Chaotic Dynamics at Homoclinic Bifurcations.* Cambridge University Press, Cambridge, New York, 1993.

[141] B. Parry and D. Sullivan. A topological invariant of flows on 1-dimensional spaces. *Topology*, 14:297–299, 1975.

[142] M. Peixoto. Structural stability on two-dimensional manifolds. *Topology*, 1:101–120, 1962.

[143] O. Perron. Über Matrizen. *Math. Ann.*, 64:248–263, 1907.

[144] R. Plykin. Sources and sinks for Axiom A diffeomorphisms. *USSR Math. Sb.*, 23:233–253, 1974.

[145] H. Poincaré. Sur les équations de la dynamique et le problème de trois corps. *Acta Math.*, 13:1–270, 1890.

[146] H. Poincaré. *Nouvelles Methodes de la Mécanique Céleste, 3 Volumes.* Gauthier-Villars, Paris, 1899.

[147] C. Pugh. The Closing Lemma. *Amer. J. Math.*, 89:956–1009, 1967.

[148] C. Pugh and M. Shub. Suspending subshifts. In C. Percelli and R. Sacksteder, editors, *Contributions to Geometry and Analysis*. London Math. Society, Johns Hopkins University Press, 1981.

[149] K. Reidemeister. *Knotentheorie*. Springer-Verlag, Berlin, 1932.

[150] J. Robbins. A structural stability theorem. *Ann. Math.*, 94:447–493, 1971.

[151] C. Robinson. Structural stability of vector fields. *Invent. Math.*, 99:154–175, 1974.

[152] C. Robinson. Homoclinic bifurcation to a transitive attractor of Lorenz type, I. *Nonlinearity*, 2:495–518, 1989.

[153] C. Robinson. *Dynamical Systems: Stability, Symbolic Dynamics, and Chaos.* CRC Press, Ann Arbor, MI, 1995.

[154] D. Rolfsen. *Knots and Links*. Publish or Perish, Berkely, CA, 1977.

[155] A. Rucklidge. Chaos in a low-order model of magnetoconvection. *Physica D*, 62:323–337, 1993.

[156] M. Rychlik. Lorenz attractors through Silnikov-type bifurcation: part I. *Ergod. Th. & Dynam. Sys.*, 10:793–821, 1989.

[157] M. Saito. On closed orbits of Morse-Smale flows on 3-manifolds. *Bull. Lond. Math. Soc.*, 23:482–486, 1990.

[158] S. Sawin. Links, quantum groups and TQFTs. *Bull. of the A.M.S.*, 33(4):413–445, 1996.

[159] L. Le Sceller, C. Letellier, and G. Gouesbet. Algebraic evaluation of linking numbers of unstable periodic orbits in chaotic attractors. *Phys. Rev. E*, 49(5):4693–4695, 1994.

[160] L. P. Shil'nikov. A case of the existence of a countable number of periodic motions. *Soviet Math. Dokl.*, 6:163–166, 1965.

[161] L. P. Shil'nikov. A contribution to the problem of the structure of an extended neighborhood of a rough equilibrium state of saddle-focus type. *Math. USSR Sbornik*, 10(1):91–102, 1970.

[162] M. Shub. *Global Stability of Dynamical Systems*. Springer-Verlag, Berlin, Heidelberg, New York, 1987.

[163] D. Singer. Stable orbits and bifurcations of maps of the interval. *SIAM J. Appl. Math.*, 35:260–267, 1978.

[164] S. Smale. Stable manifolds for differential equations and diffeomorphisms. *Ann. Scuola Normale Pisa*, 18:97–116, 1963.

[165] S. Smale. Differentiable dynamical systems. *Bull. Am. Math. Soc.*, 73:747–817, 1967.

[166] C. Sparrow. *The Lorenz Equations: Bifurcations, Chaos and Strange Attractors*, volume 41 of *Applied Mathematical Sciences*. Springer-Verlag, New York, NY, 1982.

[167] J. Stallings. On fibering certain 3-manifolds. In M. K. Fort Jr., editor, *Topology of 3-Manifolds*, pages 95–100. Prentice-Hall, 1962.

[168] M. Sullivan. Prime decomposition of knots in Lorenz-like templates. *J. Knot Thy. and Ram.*, 2(4):453–462, 1993.

[169] M. Sullivan. The prime decomposition of knotted periodic orbits in dynamical systems. *J. Knot Thy. and Ram.*, 3(1):83–120, 1994.

[170] M. Sullivan. An invariant for basic sets of Smale flows. Preprint, 1995.

[171] M. Sullivan. A zeta function for positive templates. *Top. and its Appl.*, 66(3):199–213, 1995.

[172] M. Sullivan. Visualizing Smale flows in S^3. Work in progress, 1996.

[173] D. Sumners. Untangling DNA. *Math. Intelligencer*, 12:71–80, 1990.

[174] P. Tait. On knots I, II, III. Technical report, Cambridge Univ. Press, London., 1898.

[175] F. Takens. Detecting strange attractors in turbulence. volume 898 of *Springer Lecture Notes in Mathematics*, pages 365–381, Berlin, Heidelberg, New York, 1980. Springer-Verlag.

[176] W. Thomson. On vortex motion. *Trans. R. Soc. Edin.*, 25:217–260, 1869.

[177] W. Thurston. The geometry and topology of three-manifolds. Unpublished notes, Princeton Math. Dept., 1979.

[178] W. Thurston. On the geometry and dynamics of diffeomorphisms of surfaces. *Bull. Am. Math. Soc.*, 19(2):417–431, 1988.

[179] C. Tresser. About some theorems by L. P. Shil'nikov. *Ann. Inst. H. Poincaré*, 40:441–461, 1984.

[180] N. Tufillaro. Topological organization of (low dimensional) chaos. In J. P. Nadal and P. Grassberger, editors, *From Statistical Physics to Statistical Inference and Back*. NATO ASI series C, Kluwer Academic Press, 1994.

[181] S. Ulam and J. von Neumann. On combinations of stochastic and deterministic processes. *Bull. Am. Math. Soc.*, 53:11–20, 1947.

[182] J. VanBuskirk. Positive knots have positive Conway polynomials. In D. Rolfsen, editor, *Knot Theory and Manifolds*, volume 1144 of *Lecture Notes in Mathematics*, pages 146–159. Springer-Verlag, 1985.

[183] V. Vassiliev. Cohomology of knot spaces. In V. I. Arnold, editor, *Theory of Singularities and its Applications*, pages 23–69, Providence, RI, 1990. Amer. Math. Soc.

[184] M. Wada. Closed orbits of nonsingular Morse-Smale flows on S^3. *J. Math. Soc. Japan*, 41(3):405–413, 1989.

[185] J. Wagoner. Markov partitions and $K2$. *Inst. Hautes Études Sci. Publ. Math.*, 65:91–129, 1987.

[186] J. Wagoner. Topological Markov chains, C^*-algbras and K_2. *Advances in Mathematics*, 71(2):133–185, 1988.

[187] F. Wicklin. *Dynamics near resonance in multi-frequency systems*. PhD thesis, Cornell University, 1993.

[188] S. Wiggins. *Global Bifurcations and Chaos: Analytical Methods*. Springer-Verlag, Berlin, Heidelberg, New York, 1988.

[189] S. Wiggins. *Introduction to Applied Nonlinear Dynamical Systems and Chaos*. Springer-Verlag, Berlin, Heidelberg, New York, 1990.

[190] R. Williams. The DA maps of Smale and structural stability. In *Global Analysis*, volume XIV, pages 329–334, Providence, RI, 1970. AMS.

[191] R. Williams. Classification of subshifts of finite type. *Ann. of Math.*, 98:120–153, 1973.

[192] R. Williams. Expanding attractors. *Inst. Hautes Études Sci. Publ. Math.*, 43:169–203, 1974.

[193] R. Williams. The structure of Lorenz attractors. In A. Chorin, J. Marsden, and S. Smale, editors, *Turbulence Seminar, Berkeley 1976/77*, volume 615 of *Springer Lecture Notes in Mathematics*, pages 94–116, 1977.

[194] R. Williams. The structure of Lorenz attractors. *Inst. Hautes Études Sci. Publ. Math.*, 50:73–79, 1979.

[195] R. Williams. Lorenz knots are prime. *Ergod. Th. and Dynam. Sys.*, 4:147–163, 1983.

[196] R. Williams. A new zeta funtion, natural for links. In *From Topology to Computation; Proceedings of the Smalefest*, pages 270–278, New York, NY, 1993. Springer-Verlag.

[197] E. Witten. Quantum field theory and the Jones polynomial. *Comm. Math. Phys.*, 121:355–399, 1989.

[198] J. Yorke and K. Alligood. Cascades of period-doubling bifurcations: a prerequisite for horseshoes. *Bull. Am. Math. Soc.*, 9:319–322, 1983.

[199] J. Yorke and K. Alligood. Period-doubling cascades of attractors: a prerequisite for horseshoes. *Commun. Math. Phys.*, 101:305–321, 1985.

Index

Springer
and the
environment

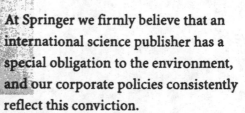

At Springer we firmly believe that an international science publisher has a special obligation to the environment, and our corporate policies consistently reflect this conviction.

We also expect our business partners – paper mills, printers, packaging manufacturers, etc. – to commit themselves to using materials and production processes that do not harm the environment. The paper in this book is made from low- or no-chlorine pulp and is acid free, in conformance with international standards for paper permanency.

Springer

Lecture Notes in Mathematics

For information about Vols. 1–1469
please contact your bookseller or Springer-Verlag

Vol. 1559: V. G. Sprindžuk, Classical Diophantine Equations. XII, 228 pages. 1993.

Vol. 1560: T. Bartsch, Topological Methods for Variational Problems with Symmetries. X, 152 pages. 1993.

Vol. 1561: I. S. Molchanov, Limit Theorems for Unions of Random Closed Sets. X, 157 pages. 1993.

Vol. 1562: G. Harder, Eisensteinkohomologie und die Konstruktion gemischter Motive. XX, 184 pages. 1993.

Vol. 1563: E. Fabes, M. Fukushima, L. Gross, C. Kenig, M. Röckner, D. W. Stroock, Dirichlet Forms. Varenna, 1992. Editors: G. Dell'Antonio, U. Mosco. VII, 245 pages. 1993.

Vol. 1564: J. Jorgenson, S. Lang, Basic Analysis of Regularized Series and Products. IX, 122 pages. 1993.

Vol. 1565: L. Boutet de Monvel, C. De Concini, C. Procesi, P. Schapira, M. Vergne. D-modules, Representation Theory, and Quantum Groups. Venezia, 1992. Editors: G. Zampieri, A. D'Agnolo. VII, 217 pages. 1993.

Vol. 1566: B. Edixhoven, J.-H. Evertse (Eds.), Diophantine Approximation and Abelian Varieties. XIII, 127 pages. 1993.

Vol. 1567: R. L. Dobrushin, S. Kusuoka, Statistical Mechanics and Fractals. VII, 98 pages. 1993.

Vol. 1568: F. Weisz, Martingale Hardy Spaces and their Application in Fourier Analysis. VIII, 217 pages. 1994.

Vol. 1569: V. Totik, Weighted Approximation with Varying Weight. VI, 117 pages. 1994.

Vol. 1570: R. deLaubenfels, Existence Families, Functional Calculi and Evolution Equations. XV, 234 pages. 1994.

Vol. 1571: S. Yu. Pilyugin, The Space of Dynamical Systems with the C^0-Topology. X, 188 pages. 1994.

Vol. 1572: L. Göttsche, Hilbert Schemes of Zero-Dimensional Subschemes of Smooth Varieties. IX, 196 pages. 1994.

Vol. 1573: V. P. Havin, N. K. Nikolski (Eds.), Linear and Complex Analysis – Problem Book 3 – Part I. XXII, 489 pages. 1994.

Vol. 1574: V. P. Havin, N. K. Nikolski (Eds.), Linear and Complex Analysis – Problem Book 3 – Part II. XXII, 507 pages. 1994.

Vol. 1575: M. Mitrea, Clifford Wavelets, Singular Integrals, and Hardy Spaces. XI, 116 pages. 1994.

Vol. 1576: K. Kitahara, Spaces of Approximating Functions with Haar-Like Conditions. X, 110 pages. 1994.

Vol. 1577: N. Obata, White Noise Calculus and Fock Space. X, 183 pages. 1994.

Vol. 1578: J. Bernstein, V. Lunts, Equivariant Sheaves and Functors. V, 139 pages. 1994.

Vol. 1579: N. Kazamaki, Continuous Exponential Martingales and BMO. VII, 91 pages. 1994.

Vol. 1580: M. Milman, Extrapolation and Optimal Decompositions with Applications to Analysis. XI, 161 pages. 1994.

Vol. 1581: D. Bakry, R. D. Gill, S. A. Molchanov, Lectures on Probability Theory. Editor: P. Bernard. VIII, 420 pages. 1994.

Vol. 1582: W. Balser, From Divergent Power Series to Analytic Functions. X, 108 pages. 1994.

Vol. 1583: J. Azéma, P. A. Meyer, M. Yor (Eds.), Séminaire de Probabilités XXVIII. VI, 334 pages. 1994.

Vol. 1584: M. Brokate, N. Kenmochi, I. Müller, J. F. Rodriguez, C. Verdi, Phase Transitions and Hysteresis. Montecatini Terme, 1993. Editor: A. Visintin. VII. 291 pages. 1994.

Vol. 1585: G. Frey (Ed.), On Artin's Conjecture for Odd 2-dimensional Representations. VIII, 148 pages. 1994.

Vol. 1586: R. Nillsen, Difference Spaces and Invariant Linear Forms. XII, 186 pages. 1994.

Vol. 1587: N. Xi, Representations of Affine Hecke Algebras. VIII, 137 pages. 1994.

Vol. 1588: C. Scheiderer, Real and Étale Cohomology. XXIV, 273 pages. 1994.

Vol. 1589: J. Bellissard, M. Degli Esposti, G. Forni, S. Graffi, S. Isola, J. N. Mather, Transition to Chaos in Classical and Quantum Mechanics. Montecatini Terme, 1991. Editor: S. Graffi. VII, 192 pages. 1994.

Vol. 1590: P. M. Soardi, Potential Theory on Infinite Networks. VIII, 187 pages. 1994.

Vol. 1591: M. Abate, G. Patrizio, Finsler Metrics – A Global Approach. IX, 180 pages. 1994.

Vol. 1592: K. W. Breitung, Asymptotic Approximations for Probability Integrals. IX, 146 pages. 1994.

Vol. 1593: J. Jorgenson & S. Lang, D. Goldfeld, Explicit Formulas for Regularized Products and Series. VIII, 154 pages. 1994.

Vol. 1594: M. Green, J. Murre, C. Voisin, Algebraic Cycles and Hodge Theory. Torino, 1993. Editors: A. Albano, F. Bardelli. VII, 275 pages. 1994.

Vol. 1595: R.D.M. Accola, Topics in the Theory of Riemann Surfaces. IX, 105 pages. 1994.

Vol. 1596: L. Heindorf, L. B. Shapiro, Nearly Projective Boolean Algebras. X, 202 pages. 1994.

Vol. 1597: B. Herzog, Kodaira-Spencer Maps in Local Algebra. XVII, 176 pages. 1994.

Vol. 1598: J. Berndt, F. Tricerri, L. Vanhecke, Generalized Heisenberg Groups and Damek-Ricci Harmonic Spaces. VIII, 125 pages. 1995.

Vol. 1599: K. Johannson, Topology and Combinatorics of 3-Manifolds. XVIII, 446 pages. 1995.

Vol. 1600: W. Narkiewicz, Polynomial Mappings. VII, 130 pages. 1995.

Vol. 1601: A. Pott, Finite Geometry and Character Theory. VII, 181 pages. 1995.

Vol. 1602: J. Winkelmann, The Classification of Three-dimensional Homogeneous Complex Manifolds. XI, 230 pages. 1995.

Vol. 1603: V. Ene, Real Functions – Current Topics. XIII, 310 pages. 1995.

Vol. 1604: A. Huber, Mixed Motives and their Realization in Derived Categories. XV, 207 pages. 1995.

Vol. 1605: L. B. Wahlbin, Superconvergence in Galerkin Finite Element Methods. XI, 166 pages. 1995.

Vol. 1606: P.-D. Liu, M. Qian, Smooth Ergodic Theory of Random Dynamical Systems. XI, 221 pages. 1995.

Vol. 1607: G. Schwarz, Hodge Decomposition – A Method for Solving Boundary Value Problems. VII, 155 pages. 1995.

Vol. 1608: P. Biane, R. Durrett, Lectures on Probability Theory. Editor: P. Bernard. VII, 210 pages. 1995.

Vol. 1609: L. Arnold, C. Jones, K. Mischaikow, G. Raugel, Dynamical Systems. Montecatini Terme, 1994. Editor: R. Johnson. VIII, 329 pages. 1995.

Vol. 1610: A. S. Üstünel, An Introduction to Analysis on Wiener Space. X, 95 pages. 1995.

Vol. 1611: N. Knarr, Translation Planes. VI, 112 pages. 1995.

Vol. 1612: W. Kühnel, Tight Polyhedral Submanifolds and Tight Triangulations. VII, 122 pages. 1995.

Vol. 1613: J. Azéma, M. Emery, P. A. Meyer, M. Yor (Eds.), Séminaire de Probabilités XXIX. VI, 326 pages. 1995.

Vol. 1614: A. Koshelev, Regularity Problem for Quasilinear Elliptic and Parabolic Systems. XXI, 255 pages. 1995.

Vol. 1615: D. B. Massey, Lê Cycles and Hypersurface Singularities. XI, 131 pages. 1995.

Vol. 1616: I. Moerdijk, Classifying Spaces and Classifying Topoi. VII, 94 pages. 1995.

Vol. 1617: V. Yurinsky, Sums and Gaussian Vectors. XI, 305 pages. 1995.

Vol. 1618: G. Pisier, Similarity Problems and Completely Bounded Maps. VII, 156 pages. 1996.

Vol. 1619: E. Landvogt, A Compactification of the Bruhat-Tits Building. VII, 152 pages. 1996.

Vol. 1620: R. Donagi, B. Dubrovin, E. Frenkel, E. Previato, Integrable Systems and Quantum Groups. Montecatini Terme, 1993. Editors:M. Francaviglia, S. Greco. VIII, 488 pages. 1996.

Vol. 1621: H. Bass, M. V. Otero-Espinar, D. N. Rockmore, C. P. L. Tresser, Cyclic Renormalization and Auto-morphism Groups of Rooted Trees. XXI, 136 pages. 1996.

Vol. 1622: E. D. Farjoun, Cellular Spaces, Null Spaces and Homotopy Localization. XIV, 199 pages. 1996.

Vol. 1623: H.P. Yap, Total Colourings of Graphs. VIII, 131 pages. 1996.

Vol. 1624: V. Brînzănescu, Holomorphic Vector Bundles over Compact Complex Surfaces. X, 170 pages. 1996.

Vol.1625: S. Lang, Topics in Cohomology of Groups. VII, 226 pages. 1996.

Vol. 1626: J. Azéma, M. Emery, M. Yor (Eds.), Séminaire de Probabilités XXX. VIII, 382 pages. 1996.

Vol. 1627: C. Graham, Th. G. Kurtz, S. Méléard, Ph. E. Protter, M. Pulvirenti, D. Talay, Probabilistic Models for Nonlinear Partial Differential Equations. Montecatini Terme, 1995. Editors: D. Talay, L. Tubaro. X, 301 pages. 1996.

Vol. 1628: P.-H. Zieschang, An Algebraic Approach to Association Schemes. XII, 189 pages. 1996.

Vol. 1629: J. D. Moore, Lectures on Seiberg-Witten Invariants. VII, 105 pages. 1996.

Vol. 1630: D. Neuenschwander, Probabilities on the Heisenberg Group: Limit Theorems and Brownian Motion. VIII, 139 pages. 1996.

Vol. 1631: K. Nishioka, Mahler Functions and Transcendence. VIII, 185 pages.1996.

Vol. 1632: A. Kushkuley, Z. Balanov, Geometric Methods in Degree Theory for Equivariant Maps. VII, 136 pages. 1996.

Vol.1633: H. Aikawa, M. Essén, Potential Theory – Selected Topics. IX, 200 pages.1996.

Vol. 1634: J. Xu, Flat Covers of Modules. IX, 161 pages. 1996.

Vol. 1635: E. Hebey, Sobolev Spaces on Riemannian Manifolds. X, 116 pages. 1996.

Vol. 1636: M. A. Marshall, Spaces of Orderings and Abstract Real Spectra. VI, 190 pages. 1996.

Vol. 1637: B. Hunt, The Geometry of some special Arithmetic Quotients. XIII, 332 pages. 1996.

Vol. 1638: P. Vanhaecke, Integrable Systems in the realm of Algebraic Geometry. VIII, 218 pages. 1996.

Vol. 1639: K. Dekimpe, Almost-Bieberbach Groups: Affine and Polynomial Structures. X, 259 pages. 1996.

Vol. 1640: G. Boillat, C. M. Dafermos, P. D. Lax, T. P. Liu, Recent Mathematical Methods in Nonlinear Wave Propagation. Montecatini Terme, 1994. Editor: T. Ruggeri. VII, 142 pages. 1996.

Vol. 1641: P. Abramenko, Twin Buildings and Applications to S-Arithmetic Groups. IX, 123 pages. 1996.

Vol. 1642: M. Puschnigg, Asymptotic Cyclic Cohomology. XXII, 138 pages. 1996.

Vol. 1643: J. Richter-Gebert, Realization Spaces of Polytopes. XI, 187 pages. 1996.

Vol. 1644: A. Adler, S. Ramanan, Moduli of Abelian Varieties. VI, 196 pages. 1996.

Vol. 1645: H. W. Broer, G. B. Huitema, M. B. Sevryuk, Quasi-Periodic Motions in Families of Dynamical Systems. XI, 195 pages. 1996.

Vol. 1646: J.-P. Demailly, T. Peternell, G. Tian, A. N. Tyurin, Transcendental Methods in Algebraic Geometry. Cetraro, 1994. Editors: F. Catanese, C. Ciliberto. VII, 257 pages. 1996.

Vol. 1647: D. Dias, P. Le Barz, Configuration Spaces over Hilbert Schemes and Applications. VII. 143 pages. 1996.

Vol. 1648: R. Dobrushin, P. Groeneboom, M. Ledoux, Lectures on Probability Theory and Statistics. Editor: P. Bernard. VIII, 300 pages. 1996.

Vol. 1649: S. Kumar, G. Laumon, U. Stuhler, Vector Bundles on Curves – New Directions. Cetraro, 1995. Editor: M. S. Narasimhan. VII, 193 pages. 1997.

Vol. 1650: J. Wildeshaus, Realizations of Polylogarithms. XI, 343 pages. 1997.

Vol. 1651: M. Drmota, R. F. Tichy, Sequences, Discrepancies and Applications. XIII, 503 pages. 1997.

Vol. 1652: S. Todorcevic, Topics in Topology. VIII, 153 pages. 1997.

Vol. 1653: R. Benedetti, C. Petronio, Branched Standard Spines of 3-manifolds. VIII, 132 pages. 1997.

Vol. 1654: R. W. Ghrist, P. J. Holmes, M. C. Sullivan, Knots and Links in Three-Dimensional Flows. X, 208 pages. 1997.